Introduction to

Thermography Principles

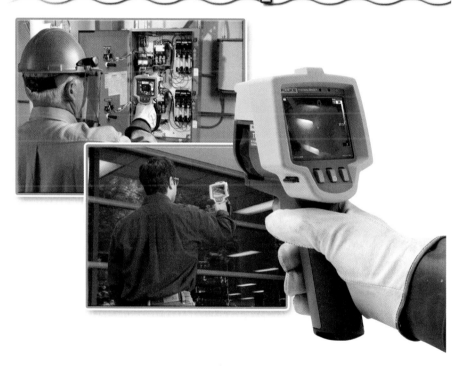

AMERICAN TECHNICAL PUBLISHERS, INC.
ORLAND PARK, ILLINOIS 60467-5756

The Snell Group™

Thank you to the authors, John Snell and Michael Stuart:

John is a long-time leader in the thermographic industry, having first used an infrared camera in 1983. In 1986, he established John Snell & Associates to better serve the needs of the infrared industry for training and consulting services. John is the founder of The Snell Group and currently works as a training specialist for the company. In 1994, John had the honor of becoming one of the first thermographers in the world to obtain a Level III Thermal/Infrared certificate from ASNT.

Michael is a practicing T/IRT Level III Thermographer, certified in compliance with ASNT standards for electrical, mechanical, and building inspection and analysis. He is also the Senior Product Manager for Thermal Imaging Products with Fluke Corporation. He conducts training for customers and Fluke personnel in the fundamentals of thermal imaging for troubleshooting, preventive maintenance, predictive maintenance, building inspection, weatherization, and energy auditing.

1 2 3 4 5 6 7 8 9 – 09 – 9 8 7

Printed in the United States of America

ISBN 978-0-8269-1535-1

 This book is printed on recycled paper.

Introduction to Thermography Principles

TABLE OF CONTENTS

INTRODUCTION

Introduction to Thermography Principles, created in cooperation with Fluke Corporation and The Snell Group, is designed to provide an introduction to thermal imager operation principles and procedures. Thermal imagers have become essential trouble-shooting and preventive maintenance discovery and diagnostic tools for electricians and technicians in industrial, process, and commercial applications. They are also a key tool for service providers to help build their businesses in the building diagnostic and inspection industries. *Introduction to Thermography Principles* covers the fundamental theory, operation, and application of using thermal imagers.

Additional information related to test instruments, troubleshooting, maintenance, and building application principles is available from Fluke Corporation at www.fluke.com/thermography, The Snell Group at www.thesnellgroup.com, and American Technical Publishers, Inc. at www.go2atp.com.

The Publisher

INTRODUCTION TO INFRARED THERMOGRAPHY AND THERMAL IMAGERS

*T*hermal imagers operate based on infrared thermography principles. A thermal imager is used as a cost-saving, and often money-making, test tool for troubleshooting, maintenance and inspection of electrical systems, mechanical systems, and building envelopes.

INFRARED THERMOGRAPHY

Infrared thermography is the science of using electronic optical devices to detect and measure radiation and correlating that to surface temperature. *Radiation* is the movement of heat that occurs as radiant energy (electromagnetic waves) moves without a direct medium of transfer. Modern infrared thermography is performed by using electronic optical devices to detect and measure radiation and correlate it to the surface temperature of the structure or equipment being inspected.

Humans have always been able to detect infrared radiation. The nerve endings in human skin can respond to temperature differences as little as $\pm 0.009°C$ $(0.005°F)$. Although extremely sensitive, human nerve endings are poorly designed for nondestructive thermal evaluation.

For example, even if humans had the thermal capabilities of animals that can find warm-blooded prey in the dark, it is probable that better heat detection tools would still be needed. Because humans have physical limitations in detecting heat, mechanical and electronic devices that are hypersensitive to heat have been developed. These devices are the standard for the thermal inspection of countless applications.

HISTORY OF INFRARED TECHNOLOGY

The derivation of "infrared" is "past red," referring to the place this wavelength holds in the spectrum of electromagnetic radiation. The term "thermography" is derived from root words meaning "temperature picture." The roots of thermography can be credited to the German astronomer Sir William Herschel who, in 1800, performed experiments with sunlight.

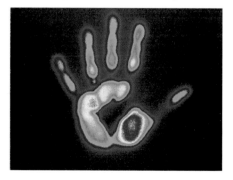

A thermal image of the residual heat transferred from a hand to the surface of a painted wall can be easily detected with a thermal imager.

Herschel discovered infrared radiation by passing sunlight through a prism and holding a thermometer in the various colors where he measured the temperature of each color using a sensitive mercury thermometer. Herschel discovered that the temperature increased when he moved out beyond red light into an area he referred to as "dark heat." "Dark Heat" was the region of the electromagnetic spectrum currently known as infrared heat and recognized as electromagnetic radiation.

Twenty years later, the German physicist Thomas Seebeck discovered the thermoelectric effect. This led to the invention of the thermomultiplier, an early version of a thermocouple, by the Italian physicist Leopoldo Nobili in 1829. This simple contact device is based on the premise that the voltage difference between two dissimilar metals changes with temperature. Nobili's partner, Macedonio Melloni, soon refined the thermomultiplier into a thermopile (an arrangement of thermomultipliers in series) and focused thermal radiation on it in such a way that he could detect body heat from a distance of 9.1 m (30').

In 1880, the American astronomer Samuel Langley, used a bolometer to detect body heat from a cow over 304 m (1000') away. Rather than measuring voltage difference, a bolometer measures the change in electrical resistance related to the change in temperature. Sir William Herschel's son, Sir John Herschel, using a device called an evaporograph, produced the first infrared image in 1840. The thermal image resulted from the differential evaporation of a thin film of oil and was viewed by light reflecting off the oil film.

A *thermal imager* is a device that detects heat patterns in the infrared-wavelength spectrum without making direct contact with equipment. **See Figure 1-1.** Early versions of thermal imagers were known as photo-conducting detectors. From 1916 through 1918, the American inventor Theodore Case experimented with photo-conducting detectors to produce a signal through direct interaction with photons rather than through heating. The result was a faster, more sensitive photo-conducting detector. During the 1940s and 1950s, thermal imaging technology was expanded to meet a growing number of military applications. German scientists discovered that by cooling the photo-conducting detector, overall performance increased.

It was not until the 1960s that thermal imaging was used for nonmilitary applications. Although early thermal imaging systems were cumbersome, slow to acquire data, and had poor resolution, they were used for industrial applications such as the inspection of large electrical transmission and distribution systems. Continued advances in the 1970s for military applications produced the first portable systems that could be used for such applications as building diagnostics and the nondestructive testing of materials.

TECH-TIP

Original versions of thermal imagers displayed thermal images through the use of black and white cathode ray tubes (CRTs). Permanent records were possible with photographs or magnetic tape.

Thermal Imagers

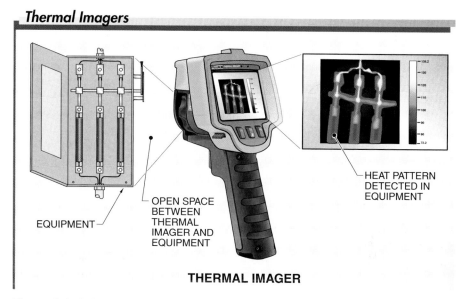

EQUIPMENT

OPEN SPACE BETWEEN THERMAL IMAGER AND EQUIPMENT

HEAT PATTERN DETECTED IN EQUIPMENT

THERMAL IMAGER

Figure 1-1. A thermal imager is a device that detects heat patterns in the infrared-wavelength spectrum without making direct contact with equipment.

Thermal imaging systems in the 1970s were durable and reliable but the quality of the images was poor compared to modern thermal imagers. By the beginning of the 1980s, thermal imaging was being widely used for medical purposes, in mainstream industry, and for building inspections. Thermal imaging systems were calibrated to produce fully radiometric images, so that radiometric temperatures could be measured anywhere in the image. A *radiometric image* is a thermal image that contains temperature measurement calculations for various points within the image.

Reliable thermal imager coolers were refined to replace the compressed or liquefied gas that had been used to cool thermal imagers. Less expensive, tube-based, pyroelectric vidicon (PEV) thermal imaging systems were also developed and widely produced. Although not radiometric, PEV thermal imaging systems were lightweight, portable, and operable without cooling.

In the late 1980s, a new device known as a focal-plane array (FPA) was released from the military into the commercial marketplace. A *focal-plane array (FPA)* is an image-sensing device consisting of an array (typically rectangular) of infrared-sensing detectors at the focal plane of a lens. **See Figure 1-2.**

This was a significant improvement over original scanned detectors and the result was an increase in image quality and spatial resolution. Typical arrays on modern thermal imagers have pixels that range from 16×16 to 640×480. A *pixel,* in this sense, is the smallest independent element of an FPA that can detect infrared energy. For specialized applications, arrays are available with pixels in excess of 1000×1000. The first number represents the number of vertical columns while the second number represents the number of horizontal rows displayed on the screen. For example, an array of 160 × 120 is equal to 19,200 total pixels (160 pixels × 120 pixels = 19,200 total pixels).

Focal Plane Arrays

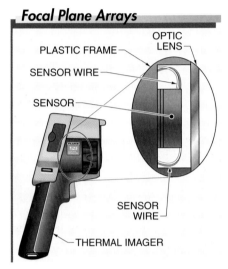

Figure 1-2. A focal-plane array (FPA) is an image-sensing device consisting of an array (typically rectangular) of light-sensing pixels at the focal plane of a lens.

The development of FPA technology utilizing various detectors has increased since the year 2000. A *long-wave thermal imager* is a thermal imager that detects infrared energy in a wavelength band that is between 8 μm and 15 μm. A *micron (μm)* is a unit of length measurement equal to one-thousandth of a millimeter (0.001 m). A *mid-wave thermal imager* is a thermal imager that detects infrared energy in a wavelength band that is between 2.5 μm and 6 μm. Both long- and mid-wave thermal imaging systems are now available in fully radiometric versions, often with image fusion and thermal sensitivities of 0.05°C (0.09°F) or less.

The cost of these systems has dropped by more than a factor of ten over the past decade and quality has dramatically increased. Furthermore, the use of computer software for image processing has grown tremendously. Nearly all commercially available, modern infrared systems utilize software to facilitate analysis and report writing. Reports can be quickly created and sent electronically over the Internet or preserved in a common format, such as a PDF, and recorded on various types of digital storage devices.

THERMAL IMAGER OPERATION

It is useful to have a general understanding of how thermal imaging systems operate because it is extremely important for a thermographer to work within the limitations of the equipment. This allows for the most accurate detection and analysis of potential problems. The purpose of a thermal imager is to detect the infrared radiation given off by the target. **See Figure 1-3.** A *target* is an object to be inspected with a thermal imager.

Infrared radiation is focused by the thermal imager's optics onto a detector resulting in a response, usually a change in voltage or electrical resistance, that is read by the electronics in the thermal imaging system. The signal produced by the thermal imager is converted into an electronic image (thermogram) on a display screen. A *thermogram* is an image of a target electronically processed onto a display screen where different color tones correspond to the distribution of infrared radiation over the surface of the target. In this simple process, the thermographer is able to view the thermogram that corresponds to the radiant energy coming off the surface of the target.

Targets

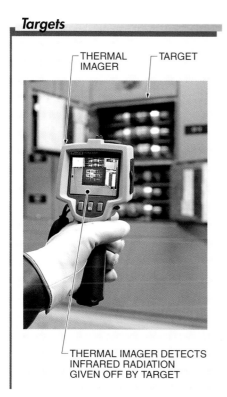

THERMAL IMAGER DETECTS
INFRARED RADIATION
GIVEN OFF BY TARGET

Figure 1-3. A target is an object to be inspected with a thermal imager. The purpose of a thermal imager is to detect the infrared radiation given off by the target.

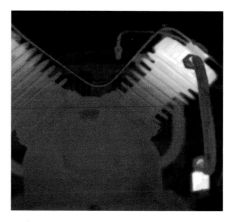

A thermogram is electronically processed onto a display screen where different color tones correspond to the distribution of infrared radiation over the surface of a target.

Lenses. Thermal imagers have at least one lens. An imager lens takes infrared radiation and focuses it on an infrared detector. The detector responds and creates an electronic (thermal) image or thermogram. The lens on a thermal imager is used to collect and focus the incoming infrared radiation on the detector. The lenses of most long-wave thermal imagers are made of germanium (Ge). Thin layers of antireflection coatings improve transmission of the lenses.

Thermal Imager Components

Typical thermal imagers consist of several common components including the lens, lens cover, display, detector and processing electronics, controls, data storage devices, and data processing and report generation software. These components can vary depending on the type and model of the thermal imaging system. **See Figure 1-4.**

TECH-TIP

Because of the ongoing need to conserve energy, municipalities and government agencies use infrared aerial scans made from adaptations to military aerial thermal maps. The purpose of these scans is to provide communities, residents, and businesses with information regarding the loss of heat in their buildings.

Thermal Imagers

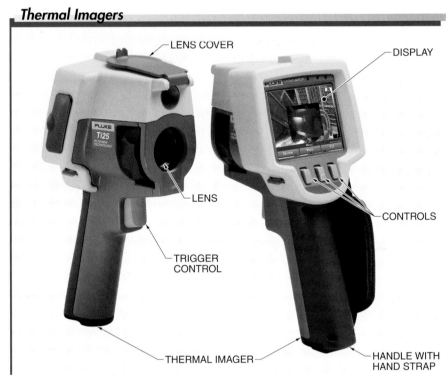

Figure 1-4. Typical thermal imagers have several common components including the lens, lens cover, display, controls, and handle with hand strap.

Thermal imagers typically have a carrying case to house the instrument, software, and other relevant equipment for field usage.

Displays. A thermal image is displayed on a liquid crystal display (LCD) view screen located on a thermal imager. The LCD view screen must be large and bright enough to be easily viewed under the diverse lighting conditions encountered in various field locations. A display will also often provide information such as battery charge, date, time, target temperature (in °F, °C, or °K), visible light image, and a color spectrum key related to the temperature. **See Figure 1-5.**

Detector and Processing Electronics. Detector and processing electronics are used to process infrared energy into usable information. Thermal radiation from the target is focused on the detector, which is typically an electronic semiconductor material. The thermal radiation produces a measurable response from the detector. This response is processed electronically within the thermal imager to produce a thermal image on the display screen of the thermal imager.

Displays

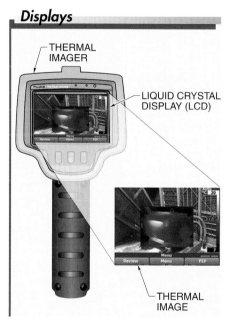

Figure 1-5. A thermal image is displayed on a liquid crystal display (LCD) located on the thermal imager.

Controls. Various electronic adjustments can be made with controls to refine a thermal image on the display. Electronic adjustments can be made to variables such as temperature range, thermal span and level, color palettes, and image fusion. Adjustments can often also be made to the emissivity and reflected background temperature. **See Figure 1-6.**

Controls

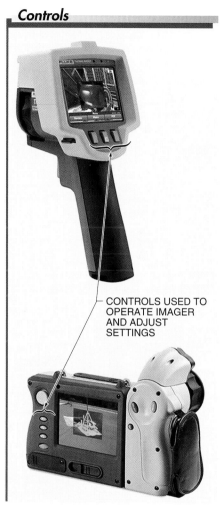

Figure 1-6. With controls, adjustments can be made to important variables such as temperature range, thermal span and level, and other settings.

Data Storage Devices. Electronic digital files containing thermal images and associated data are stored on different types of electronic memory cards or storage and transfer devices. Many infrared imaging systems also allow for storage of supplementary voice or text data as well as a corresponding visual image acquired with an integrated visual light camera.

Data Processing and Report Generation Software. The software used with most modern thermal imaging systems is both powerful and user friendly. Digital thermal and visual images are imported into a personal computer where they can be displayed using various color palettes, and where further adjustments can be made to all radiometric parameters and analysis functions. The processed images are then inserted into report templates, and either sent to a printer, stored electronically, or sent to customers via an Internet connection.

THERMOGRAPHY AND RETURN ON INVESTMENT

*T*hermography, through the use of thermal imagers, can be used to perform many critical functions for commercial and industrial environments including the troubleshooting and maintenance of equipment and the inspections of a building envelope. Thermal imagers have traditionally been considered expensive. However, the costs associated with the maintenance and unplanned downtime of a facility's operations can be greatly reduced when thermal imagers are used to perform preventive and predictive maintenance tasks.

TROUBLESHOOTING

Infrared thermography has an important function when troubleshooting problems in commercial and industrial operations. Questions about equipment condition are often raised by some abnormal condition or indication. On an obvious level, this may be as simple as a noticeable vibration, sound, or temperature reading. On a subtle level, the root cause of the problem may be difficult or impossible to discern.

A *thermal signature* is a false color picture of the infrared energy, or heat, being emitted from an object. Comparing the thermal signature of a normally operating piece of equipment to the one being evaluated for abnormal conditions, offers an excellent means of troubleshooting. **See Figure 2-1.** The primary benefits of infrared thermography are that tests can be performed quickly and without destruction to equipment. Also, since thermal imagers do not require contact, they can be used while the equipment or component is in operation.

Even if an abnormal thermal image is not fully interpreted by a thermographer, it can be used to determine if further

testing may be required. For example, it is easy to perform a quick inspection of an electric motor and understand if there are irregularities with the bearings and any coupling. A motor bearing that is significantly warmer than the motor casing suggests the possibility of either a lubrication or an alignment problem. An alignment problem can also be indicated if one side of the coupling is warmer than the opposite side. **See Figure 2-2.**

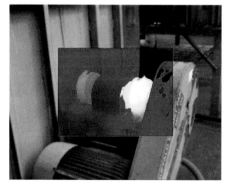

A hot bearing cap is a sign of a potential problem with either the alignment, lubrication, or issues with the motor or equipment it is connected to.

Thermal Signatures

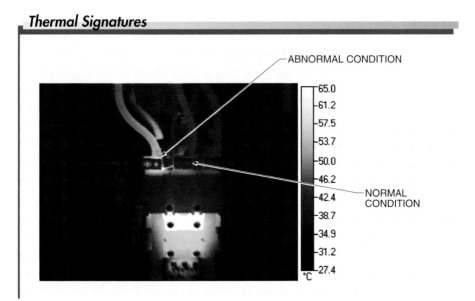

Figure 2-1. The thermal signatures of operating equipment can quickly indicate normal and abnormal conditions.

Troubleshooting Motor Bearings

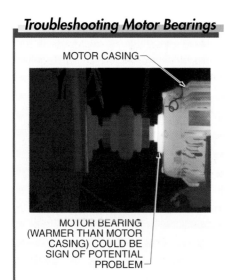

Figure 2-2. A motor bearing that is significantly warmer than the motor casing suggests the possibility of either a lubrication or alignment problem.

The key to successful troubleshooting using thermal imaging is having a good understanding of the basic requirements needed to detect potential problems or abnormal conditions in any specific piece of equipment when they are present. For example, using a thermal imager to trouble-shoot an electrical disconnect switch when it is not energized has no value because potential problems (hot spots) will not be visible unless the electrical disconnect switch is energized. Likewise, to success-fully troubleshoot a steam trap, it must be observed as it goes through a complete operating cycle.

Knowing exactly what conditions are needed for troubleshooting a particular piece of equipment is not always simple. Along with thermographer experience, a solid understanding of variables such as heat transfer, radiometry, camera use, and equipment function and failure are all required for successful troubleshooting. *Radiometry* is the detection and measurement of radiant electromagnetic energy, specifically in the infrared part of the spectrum.

Preventive Maintenance

Preventive maintenance (PM) is scheduled work required to maintain equipment in peak operating condition. PM minimizes equipment malfunctions and failures while maintaining optimum production efficiency and safety conditions in the facility. This results in increased service life, reduced downtime, and greater overall plant efficiency. PM tasks and their frequency for each piece of equipment are determined by manufacturer's specifications, equipment manuals, trade publications, and worker experience.

A strategy of providing a comprehensive understanding of the operating condition of equipment through condition-based assessment and monitoring is considered critical to PM programs. PM programs that include condition-based assessment and monitoring of equipment are performed more easily through the use of thermal imaging equipment. By reviewing thermal images of equipment, repair/replace decisions become more effective, overall costs are reduced, and equipment reliability is increased. When production requires that a piece of equipment be completely functional, production management can be assured it will be ready to perform the job as intended.

Maintenance is a sophisticated set of activities driven by specific methods. In recent years, it has been discovered that many of the old methods, via preventive maintenance, often caused more problems than they solved. Furthermore, they did not always have a particularly good return on the investment.

Predictive Maintenance

Predictive maintenance (PdM) is the monitoring of wear conditions and equipment characteristics against a predetermined tolerance to predict possible malfunctions or failures. Equipment-operation data is gathered and analyzed to show trends in performance and component characteristics. Repairs are made as required.

PdM often requires a substantial investment in monitoring equipment and training for personnel. It is most commonly used on expensive or critical operating equipment. Data collected from monitoring equipment is analyzed on a regular basis to determine if values are within acceptable tolerances. **See Figure 2-3.** Maintenance procedures are performed if values are outside acceptable tolerances. The equipment is then closely monitored after maintenance procedures are performed. If the problem recurs, the equipment application and design are analyzed and changes are made as required.

With a successful PdM program, preventive maintenance can usually be reduced. Certain maintenance tasks, such as lubrication or cleaning, are performed when they are actually needed rather than on a fixed schedule. Thermography and thermal imaging can be used to determine equipment health and, when the condition is in question, it is also used to monitor equipment until a period for maintenance is available.

An acceptance inspection is an inspection performed at the time of initial equipment installation, or replacement of a component, in order to establish a baseline condition of that equipment. The baseline condition is used for verification of manufacturer performance specifications or comparisons at later points in time. Acceptance inspections of new or rebuilt equipment are critical to cost-effective PdM programs.

Whether installing a new motor-control center, roof, steam line, or building insulation, thermal imaging is used to document the actual equipment condition at the time of acceptance. A thermal image can be used to determine if the installation was properly performed. If a deficiency in the installation is discovered, it can be immediately corrected or, as circumstances allow, monitored until a repair period can be scheduled.

Regardless of the maintenance programs used within a company, the use of thermography and thermal imagers is beneficial. When used for troubleshooting and maintenance, the advantages are reduced equipment outages and increased operating time. Other major benefits include large returns on investments in reliability maintenance, cost savings through reduced hours, and the reduced overall frustration of maintenance technicians.

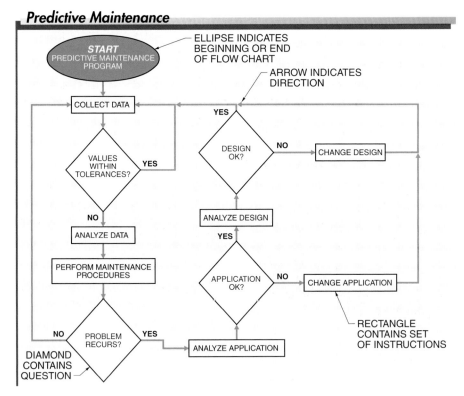

Figure 2-3. Predictive maintenance is most commonly used on the expensive or critical equipment in a facility.

TRAINING AND SAFETY

*T*hermal imagers can be used to perform a wide variety of tasks in commercial and industrial environments. Many of these tasks can take place in areas where exposure to hazards, such as energized electrical equipment and heights, are common. Proper training in thermal imager use along with implementation of safety rules is necessary to safely and efficiently perform the required tasks. Various written standards and procedures are used for proper training.

THERMOGRAPHER QUALIFICATION AND CERTIFICATION

Learning to use a modern thermal imager is relatively easy. It can typically be mastered with basic training and hands-on practice. However, properly interpreting a thermal image is often more difficult. It requires not only a background in the application of thermography but also additional, more extensive, training and hands-on experience with thermal imagers.

To gain full return on investment in thermography, it is important to qualify and certify thermographers. Regardless of the specific use of the technology, thermographer qualification is based on training, experience, and testing in one of three categories of certification. **See Figure 3-1.**

While thermographer certification represents an investment, it is an investment that typically pays large returns. Not only do certified personnel produce higher quality inspections, their inspections are also more technically

consistent. Uncertified thermographers are more likely to make costly and dangerous mistakes. These mistakes often result in serious consequences, such as inaccurate recommendations regarding the criticality of the problems discovered or problems being completely missed altogether. While the appropriate qualification is important, written inspection procedures are also important for attaining high-quality results.

Thermographer Certification Levels	
LEVEL 1	Qualified to gather high-quality data and sort the data based on written pass/fail criteria.
LEVEL 2	Qualified to set up and calibrate equipment, interpret data, create reports and supervise Level 1 personnel.
LEVEL 3	Qualified to develop inspection procedures, interpret relevant codes, and manage a program including overseeing or providing training and testing.

Figure 3-1. There are three levels of thermographer certification.

13

In the United States, certification is issued by the employer in compliance with the standards of the American Society for Nondestructive Testing. The *American Society for Nondestructive Testing (ASNT)* is an organization that helps create a safer environment by serving the nondestructive testing professions and promoting nondestructive testing technologies through publishing, certification, research, and conferencing. In other parts of the world, certification is provided by a central certifying body in each country that complies with the standards of the International Organization for Standardization. The *International Organization for Standardization (ISO)* is a nongovernmental, international organization that is comprised of national standards institutions from more than 90 countries.

Under both models, qualification is based upon the appropriate training, as outlined in the documents of the relevant standards. A period of qualifying experience and some form of written and hands-on examination are also required.

TECH-TIP

Prior to performing a thermal inspection, the thermographer should perform a "walk-down" of the planned inspection route to ensure efficiency and to look for possible safety concerns.

SAFETY IN THE WORKPLACE

A portion of any certification program is the awareness of the inherent dangers of thermography and the techniques and skills needed to ensure safety in the workplace. Common sense dictates much of what constitutes safe work practice but special precautions often apply to a specific application. For example, thermographers who inspect electrical systems may have a greater exposure to the potential of an arc blast.

In many instances, they are inspecting energized equipment which, immediately after the enclosure has been opened, can trigger a phase-to-phase or phase-to-ground arc. An *arc flash* is an extremely high-temperature discharge produced by an electrical fault in the air. Arc flash temperatures can reach 35,000°F (19,427°C).

An *arc blast* is an explosion that occurs when air surrounding electrical equipment becomes ionized and conductive. The threat of an arc blast is greatest for electrical systems of 480 V and higher.

A *flash protection boundary* is the distance at which personal protective equipment (PPE) is required for the prevention of burns when an arc flash occurs. **See Figure 3-2.** While a circuit that is being repaired should always be de-energized, the possibility exists that nearby circuits are still energized within the flash protection boundary. Therefore, barriers, such as insulation blankets, along with the proper PPE must be used to protect against an arc flash. However, the consequences of an arc blast are often deadly and extensive. Safety must always be practiced.

While the risk for an arc blast is minimized by not opening the cover or door to an enclosure this also eliminates most of the benefit of thermography, as we cannot see through the enclosure covers. **See Figure 3-3.** However, many enclosures are now installed with special infrared transparent windows or viewports. These features can reduce the risk of arcing and yield good results.

Flash Protection Boundaries

Nominal System (Voltage, Range, Phase to Phase*)	Limited Approach Boundary		Restriced Approach Boundary (Allowing for Accidental Movement)	Prohibited Approach Boundary
	Exposed Movable Conductor	Exposed Fixed-Circuit Part		
0 to 50	N/A	N/A	N/A	N/A
51 to 300	10'-0"	3'-6"	Avoid Contact	Avoid Contact
301 to 750	10'-0"	3'-6"	1'-0"	0'-1"
751 to 15,000	10'-0"	5'-0"	2'-2"	0'-7"

* in V

Figure 3-2. A flash protection boundary is the suggested distance at which PPE is required for the prevention of burns when an arc flash occurs.

Electrical Enclosures

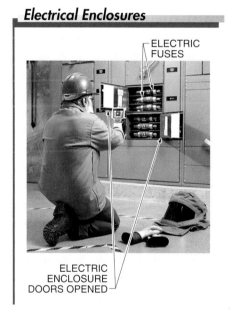

ELECTRIC FUSES

ELECTRIC ENCLOSURE DOORS OPENED

Figure 3-3. When electrical enclosures must be opened, procedures should be carefully developed, implemented, and followed that will minimize the risk of an arc flash.

When enclosures must be opened, procedures should be carefully developed, implemented, and followed that will minimize the risk of an arc flash. National Fire Protection Agency (NFPA) 70E, is one of several standards that can be useful when developing such procedures.

Routine electrical inspection can be much safer and more effective when conducted by a team. The team could consist of two people such as the thermographer and the qualified person who opens enclosures, measures loads, and safely closes the enclosure once work has been completed. A *qualified person* is a person who has knowledge and skills related to the construction and operation of electrical equipment and has received appropriate safety training.

Work for building inspections is typically less risky. However, risks do exist such as when accessing crawlspaces and attics. Care must also be taken when being exposed to construction work that is in progress.

Thermographers working in any industrial environment must always be aware of other hazards including the potential for trips and falls and enclosed-space entry hazards. Bright clothing may also be required in many environments. On roofs, precaution must be taken for fall hazards, not only at the roof edge but also for simple changes in elevation or over a structurally weakened roof deck. Work performed on roofs should never be performed alone.

Furthermore, special precautions must be taken at night. A thermographer can become night blind when viewing a thermal image in the bright display of an imaging system. *Night blindness* is a condition that occurs when a thermographer's eyes adjust to view a brightly lit display screen and, as a result, are not adjusted to see a dark object.

Accidents typically occur when work is not planned or when the nature of the planned work changes but the plan does not. A safe-work plan should always be developed and followed. When circumstances change, the plan must be reevaluated for any necessary changes.

The *Occupational Safety and Health Administration (OSHA)* is a United States government agency established under the Occupational Safety and Health Act of 1970 that requires employers to provide a safe environment for their employees. For example, OSHA requires that work areas must be free from hazards that are likely to cause serious harm. OSHA provisions are enforced by the U.S. government and safe-work plans can be developed within OSHA guidelines.

Thermal inspections of high-power electrical equipment must be performed at a safe distance from the equipment.

STANDARDS AND WRITTEN INSPECTION PROCEDURES

Written inspection procedures are essential to produce high-quality results. For example, trying to bake a cake without a recipe would be much more difficult than having a recipe to follow. Written inspection procedures can be referred to as "recipes for success."

Creating these "recipes for success," while an investment, does not need to be difficult. Typically, it is useful to involve a small group of individuals who have relevant experience in the inspection process in order to represent different viewpoints, areas of expertise, and responsibilities. Once a written inspection procedure has been developed, it should be tested thoroughly and periodically reviewed by certified personnel to ensure that it continues to represent best practices.

Many inspection standards exist that can provide a foundation for simple written inspection procedures. For example, committees of professionals have worked with both the ISO and American Society of Testing Materials (ASTM) International to develop a number of relevant standards. *American Society of Testing Materials (ASTM) International* is a technical society and primary developer of voluntary standards, related technical information, and services that promote public health and safety. ASTM International also contributes to the reliability of products, materials, and services.

These standards help to determine the performance of infrared systems. They also describe the best practice for inspections of building insulation, air leakage, electrical and mechanical systems, roofs, and highway bridge decks. Other standards organizations in individual countries may have additional standards that can be used. For example, many have standards governing electrical safety that will apply directly to the work of thermographers inspecting electrical systems.

Due to the large variety of thermal imagers available today and the wide range of prices, infrared technology has become very easy to access. However, organizations that invest in the development of solid thermal imaging programs with inspection procedures and qualified personnel have a distinct advantage. Typically, they will have long-term benefits that other organizations may not receive. **See Figure 3-4.**

Thermal Imagers

FOR GENERAL MAINTENANCE, TROUBLESHOOTING, AND BASIC INSPECTIONS

FOR SPECIALIZED, COMPLEX, OR INTENSIVE APPLICATIONS REQUIRING ENHANCED DETECTION AND ANALYSIS CAPABILITIES

Figure 3-4. There are different thermal imagers available for different types of applications and inspections.

PRACTICAL APPLIED THEORY

Thermodynamic theory and science is based on the variations of heat transfer between different materials. Thermal imagers take readings based on the principles of basic thermodynamics. Technicians must be able to understand the limitations of thermography and thermal imagers when taking readings of various structures, equipment, and materials.

BASIC THERMODYNAMICS

Thermodynamics is the science of how thermal energy (heat) moves, transforms, and affects all matter. To use modern infrared equipment, it is essential to understand the basics of both heat transfer and radiation physics. As powerful as modern equipment is, it still cannot think for itself. The value of modern equipment is defined by a thermographer's ability to interpret data, which requires a practical understanding of the basics of heat transfer and radiation physics.

Energy is the capacity to do work. Energy can take various forms. For example, a coal-fired electrical generation plant converts the chemical energy from fossil fuel to thermal energy through combustion. That, in turn, produces mechanical energy or motion in a turbine generator that is then converted to electrical energy. During these conversions, although the energy does become more difficult to harness, none of it is lost.

The *first law of thermodynamics* is a scientific law that states when mechanical work is transformed into heat, or when heat is transformed into work, the amount of work and heat are always equivalent. An advantage for thermographers is the fact that a byproduct of nearly all energy conversions is heat, or thermal energy. Energy cannot be created or destroyed, only altered.

Temperature is a measure of the relative hotness or coldness of a body when compared to another. We unconsciously make comparisons to our body temperature or ambient air temperature and to the boiling and freezing points of water.

The *second law of thermodynamics* states when a temperature difference exists between two objects, thermal energy transfers from the warmer areas (higher energy) to the cooler areas (lower energy) until thermal equilibrium is reached. A transfer of heat results in either electron transfer or increased atomic or molecular vibration. This is important because these effects are measured when temperature is measured.

METHODS OF HEAT TRANSFER

Heat energy can be transferred by any of three methods: conduction, convection, or radiation. Each method is described as either steady state or transient. During a steady state transfer, the transfer rate is constant and in the same direction over time. For example, a fully warmed up machine under constant load transfers heat at a steady state rate to its surroundings. In reality, there is no such thing as perfect steady-state heat flow. There are always small transient fluctuations but for practical purposes, they are typically ignored.

Conduction is the transfer of thermal energy from one object to another through direct contact. *Convection* is the transfer of heat that occurs when molecules move and/or currents circulate between the warm and cool regions of air, gas, or fluid. *Radiation* is the movement of heat that occurs as radiant energy (electromagnetic waves) moves without a direct medium of transfer. When a machine warms up or cools down, heat is transferred in a transient manner. Understanding these relationships is important to thermographers because the movement of heat is often closely related to the temperature of an object.

Concept of Thermal Capacitance

Thermal capacitance is the ability of a material to absorb and store heat. When heat is transferred at varying rates and/or in different directions, it is said to be transient.

Additionally, as various materials are in transition, differing amounts of energy are exchanged as they change temperature. For example, very little energy is required to change the temperature of the air in a room compared to the amount needed to change the temperature of an equal volume of water in a swimming pool. Thermal capacitance describes how much energy is added or removed for a material to change temperature. How quickly or slowly that change happens also depends on how the heat moves.

While thermal capacitance, which is what we call the relationships between heat and temperature, can be confusing, it can also be beneficial to a thermographer. For example, finding the liquid level in a tank is possible because of the difference between the thermal capacitance of the air and the liquid. When the tank is in transition, the two materials often exist at different temperatures.

Conduction

Conduction is the transfer of thermal energy from one object to another through direct contact. Heat transfer by conduction occurs primarily in solids, and to some extent in fluids, as warmer molecules transfer their energy directly to cooler, adjacent ones. For example, conduction is experienced when touching a warm mug of coffee or a cold can of soft drink.

The rate at which heat transfer occurs depends on the conductivity of the materials and the temperature difference (ΔT or delta-temperature) between the objects. These simple relationships are described more formally by Fourier's law. For example, when picking up a hot coffee cup while wearing gloves, very little heat is exchanged compared to doing so with a bare hand. A warm cup of coffee does not transfer as much energy as does a hot one because the temperature difference is not as great. Similarly, when energy is transferred at the same rate but over a larger area, more energy is transferred.

A *conductor* is a material that readily transfers heat. Metals are typically very conductive of heat. However, even the conductivity of metals can vary depending on the type of metal. For example, iron is much less conductive than aluminum. An *insulator* is a material that is inefficient at transferring heat. Materials that are inefficient conductors are known as insulators. Often these are simply materials, such as foam insulation or layered clothing, that trap small pockets of air and slow the transfer of energy. **See Figure 4-1.**

Convection

Convection is the transfer of heat that occurs when currents circulate between warm and cool regions of fluids. Convection occurs in both liquids and gases, and involves the mass movement of molecules at different temperatures. For example, a thundercloud is convection that occurs on a large scale because as masses of warm air rise, cool air sinks. On a smaller scale convection occurs as the cold cream, poured into a cup of hot coffee, sinks to the bottom of the cup.

Convective heat transfer is also determined in part by area and temperature difference. For example, an engine radiator with a large engine transfers more heat than a small engine because of its larger area. Other factors also affect convective heat transfer including the velocity of a fluid, direction of fluid flow, and surface condition of an object. An engine radiator that is blocked by dust does not transfer heat as efficiently as a clean one. As with conduction, most of us have a good practical sense of these relationships, which are more formally described by Newton's law of cooling. Natural convection occurs when warmer fluids rise and cooler fluids sink, such as in the cooling tubes of oil-filled transformers. **See Figure 4-2.**

Insulators

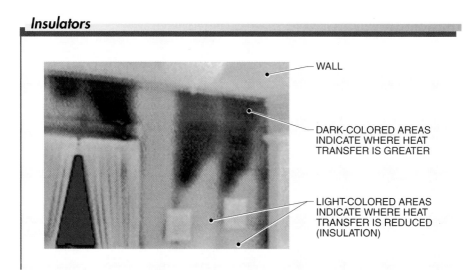

WALL

DARK-COLORED AREAS INDICATE WHERE HEAT TRANSFER IS GREATER

LIGHT-COLORED AREAS INDICATE WHERE HEAT TRANSFER IS REDUCED (INSULATION)

Figure 4-1. Insulators are installed in walls to control the transfer of heat. Poorly installed insulation does not control heat transfer adequately.

Natural Convection

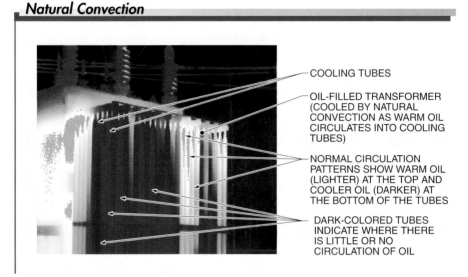

COOLING TUBES

OIL-FILLED TRANSFORMER (COOLED BY NATURAL CONVECTION AS WARM OIL CIRCULATES INTO COOLING TUBES)

NORMAL CIRCULATION PATTERNS SHOW WARM OIL (LIGHTER) AT THE TOP AND COOLER OIL (DARKER) AT THE BOTTOM OF THE TUBES

DARK-COLORED TUBES INDICATE WHERE THERE IS LITTLE OR NO CIRCULATION OF OIL

Figure 4-2. Natural convection occurs when warm oil rises and cool oil sinks, such as in the cooling tubes of an oil-filled transformer.

When convection is forced, such as with a pump or a fan, the natural relationships are generally overwhelmed because forced convection can be quite powerful. When the wind blows, we feel colder, which is evidence that we are losing heat at a faster rate than when the wind is not blowing. The wind also strongly influences the temperature of the objects inspected with thermal imaging systems.

Radiation

Radiation is the transfer of energy, including heat, that occurs at the speed of light between objects by electromagnetic energy. Because no transfer medium is required, radiation can even take place in a vacuum. An example of electromagnetic energy is feeling the heat of the sun on a cool day.

Electromagnetic energy is radiation in the form of waves with electric and magnetic properties. Electromagnetic energy can take on several forms including light, radio waves, and infrared radiation. The primary difference among all of these forms is their wavelength. While normal eyesight detects wavelengths known as visible light, thermal imagers detect wavelengths known as radiated heat (or infrared radiation). Each wavelength is on a different part of the electromagnetic spectrum.

The Stefan-Boltzmann equation describes the relationships for how heat moves as radiation. All objects radiate heat. As with conduction and convection, the net amount of energy radiated depends on area and temperature differences. The warmer an object, the more energy it radiates. For example, when a stove burner gets hotter, it radiates more energy than when it is cool.

Thermal radiation is the transmission of heat by means of electromagnetic waves. The primary difference among the waves is their wavelength. While electromagnetic radiation (light) is visible to the human eye, radiated heat is visible only through thermal imaging systems. The *electromagnetic spectrum* is the range of all types of electromagnetic radiation, based on wavelength. **See Figure 4-3.**

Electromagnetic Spectrum

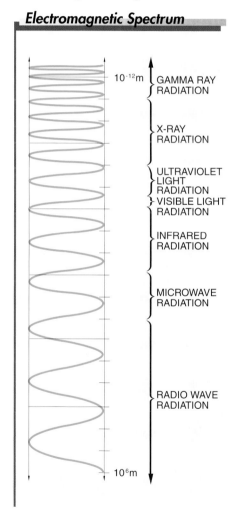

10^{-12}m	GAMMA RAY RADIATION
	X-RAY RADIATION
	ULTRAVIOLET LIGHT RADIATION
	VISIBLE LIGHT RADIATION
	INFRARED RADIATION
	MICROWAVE RADIATION
	RADIO WAVE RADIATION
10^{6}m	

Figure 4-3. The electromagnetic spectrum is the range of all types of electromagnetic radiation, based on wavelength.

Concept of Conservation of Energy

Light and infrared radiation behave similarly when they interact with various materials. Infrared radiation is reflected by some types of surfaces, such as the metal liner under a stove burner. Reflections of both warm and cool objects can be seen with infrared imagers in some surfaces, such as bright metals, which we call "thermal mirrors." In a few cases, infrared radiation is transmitted through a surface, such as through the lens of an infrared imaging system. Infrared radiation can also be absorbed by a surface, such as a hand near a hot stove burner. In this case a temperature change results, causing the surface to emit more energy.

Transmission is the passage of radiant energy through a material or structure. Infrared radiation can also be absorbed in a surface, causing the temperature to change and an emission of more energy from the surface of the object. *Absorption* is the interception of radiant energy. *Emission* is the discharge of radiant energy. Although an infrared thermal imaging system can read reflected, transmitted, absorbed, and emitted radiation, only energy that is absorbed or emitted affects the surface temperature. **See Figure 4-4.**

TECH-TIP

The roughness of a surface determines the type and direction of radiation reflection. A smooth surface is known as a specular reflector while a rough or patterned surface is known as a diffuse reflector.

Reflection, Transmission, Absorption, and Emission

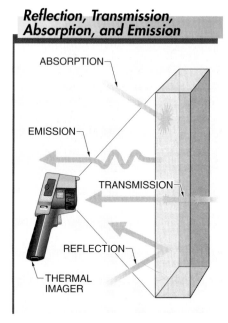

Figure 4-4. Radiation can be reflected, transmitted, absorbed, or emitted.

In addition, the quantity of heat radiated by a surface is determined by how efficiently the surface emits energy. Most nonmetallic materials, such as painted surfaces or human skin, efficiently emit energy. This means that as their temperature increases, they radiate a great deal more energy, such as with a stove burner.

Other materials, mostly metals that are unpainted or not heavily oxidized, are less efficient at radiating energy. When a bare metal surface is heated, there is comparatively little increase in the radiant heat transfer, and it is difficult to see the difference between a cool metal surface and a warm one either with our eyes or a thermal imaging system. Bare metals typically have a low emissivity (low efficiency of emission). Emissivity is characterized as a value between 0.0 and 1.0. A surface with a value of 0.10, typical for shiny copper, emits little energy compared to human skin with an emissivity of 0.98.

One of the challenges of using a thermal imager is that these instruments display energy that is normally invisible to human eyes. Sometimes this can be confusing. Not only do low-emissivity surfaces, such as metals, emit energy inefficiently, they are also reflective of their thermal surroundings. When a surface is read with a thermal imaging system, it shows a combination of emitted and reflected infrared radiation in the image. In order to make sense of what is displayed, the thermographer must understand what energy is being emitted and what energy is being reflected.

Several other factors can affect material emissivity. Besides material type, emissivity can also vary with surface condition, temperature, and wavelength. The effective emissivity of an object can also vary with the angle of view. **See Figure 4-5.**

It is not difficult to characterize the emissivity of most materials that are not shiny metals. Many materials have already been characterized and their emissivity values can be found in emissivity tables. Emissivity values should be used only as a guide. Since the exact emissivity of a material may vary from these values, skilled thermographers also need to understand how to measure the actual value. **See Figure 4-6.**

Cavities, gaps, and holes emit thermal energy at a higher rate than the surfaces around them. The same is true for visible light. The pupil of the human eye is black because it is a cavity, and the light that enters it is absorbed. When all light is absorbed by a surface we say it is "black." The emissivity of a cavity will approach 0.98 when it is seven times deeper than it is wide.

Emissivity

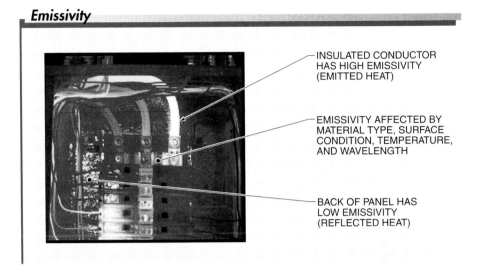

INSULATED CONDUCTOR
HAS HIGH EMISSIVITY
(EMITTED HEAT)

EMISSIVITY AFFECTED BY
MATERIAL TYPE, SURFACE
CONDITION, TEMPERATURE,
AND WAVELENGTH

BACK OF PANEL HAS
LOW EMISSIVITY
(REFLECTED HEAT)

Figure 4-5. Emissivity can be affected by material type, surface condition, temperature, and wavelength.

Emissivity Values of Common Materials

Material	Emissivity*
Aluminum, polished	0.05
Brick, common	0.85
Brick, refractory, rough	0.94
Cast iron, rough casting	0.81
Concrete	0.54
Copper, polished	0.01
Copper, oxidized to black	0.88
Electrical tape, black plastic	0.95
Glass	0.92
Lacquer, Bakelite	0.93
Paint, average oil-based	0.94
Paper, black, dull	0.94
Porcelain, glazed	0.92
Rubber	0.93
Steel, galvanized	0.28
Steel, oxidized strongly	0.88
Tar paper	0.92
Water	0.98

* Emissivities of almost all materials are measured at 0°C (32°F) but do not differ significantly at room temperature.

Figure 4-6. The emissivity values of many common materials can be found in emissivity tables.

Surface Temperature

Typically, because only the surface-temperature patterns of most objects are seen (as they are opaque) thermographers must interpret and analyze these patterns and relate them to the object's internal temperatures and structures. For example, the exterior wall of a house will display patterns of various temperatures, but the task of a thermographer is to relate them to the structure and thermal performance of the house. To accurately do this, there must be a basic understanding of how heat travels through different components and materials in the wall.

During cold weather, heat from inside the house travels through the structure of the wall to the exterior surface, and then the surface comes into thermal equilibrium with its surroundings. It is at this point that thermographers view that surface with a thermal imager and must interpret what is shown. These relationships can often be quite complex, but are best understood in many cases by simply using common sense and paying attention to basic science.

Emissivity

Metals that are unpainted or not heavily oxidized are difficult to read in a thermal image because they emit little and reflect a great deal. Whether we are just looking at the thermal patterns or actually making a radiometric temperature measurement, we need to take these factors into account. In many thermal imagers, corrections can be made for both the emissivity and reflected thermal background. Emissivity correction tables have been developed for many materials.

While emissivity correction tables can be useful for understanding how a material will behave, the reality of making a correction for most low-emissivity surfaces is that errors can be unacceptably large. Low-emissivity surfaces should be altered by some means, such as covering it with electrical tape or paint, in order to increase the emissivity. This makes both interpretation and measurement accurate and practical.

TEMPERATURE MEASUREMENT ACCURACY

The accuracy of modern infrared test instruments is quite high. When viewing high-emissivity, moderately warm surfaces within the measurement resolution of a system, test accuracy is typically ±2°C (3.6°F) or 2% of the measurement (but can vary according to the thermal imager model). Also, since infrared test instruments do not require contact with the objects being tested, infrared technol-

ogy has tremendous value because of the increased accuracy of measurements.

Because temperature measurements are based on the detection of infrared radiation, the following factors can be expected to reduce temperature measurement accuracy:

- Emissivity values below 0.6
- Temperature variations of ±30°C (54°F)
- Making measurements beyond the resolution of the system (target too small or far away)
- Field of view

Field of View (FOV)

A *field of view (FOV)* is a specification that defines the size of what is seen in the thermal image. The lens has the greatest influence on what the FOV will be, regardless of the size of the array. Large arrays, however, provide greater detail, regardless of the lens used, compared to narrow arrays. For some applications, such as work in outdoor substations or inside a building, a large FOV is useful. While smaller arrays may provide sufficient detail in a building, more detail is important in substation work. **See Figure 4-7.**

Instantaneous Field of View (IFOV)

An *instantaneous field of view (IFOV)* is a specification used to describe the capability of a thermal imager to resolve spatial detail (spatial resolution). The IFOV is typically specified as an angle in milliradians (mRad). When projected from the detector through the lens, the IFOV gives the size of an object that can be seen at a given distance.

Field of View (FOV)

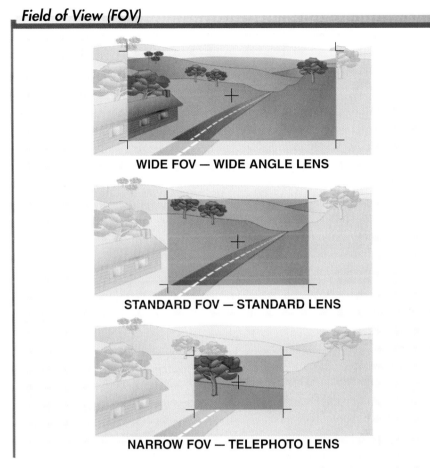

WIDE FOV — WIDE ANGLE LENS

STANDARD FOV — STANDARD LENS

NARROW FOV — TELEPHOTO LENS

Figure 4-7. The field of view (FOV) is a specification that defines the area that is seen in the thermal image when using a specific lens.

An *IFOV measurement* is the measurement resolution of a thermal imager that describes the smallest size object that can be measured at a given distance. **See Figure 4-8.** It is specified as an angle (in mRad) but is typically larger by a factor of three than the IFOV. This is due to the fact that the imager requires more information about the radiation of a target to measure it than it does to detect it. It is vital to understand and work within the spatial and measurement resolution specific to each system. Failure to do so can lead to inaccurate data or overlooked findings.

TECH-TIP

All thermal imager targets radiate energy measurable on the infrared spectrum. As the target heats up, it radiates more energy. Very hot targets radiate enough energy to be seen by the human eye.

Environmental Effects

The value of a surface measurement, even if accurate, may decrease significantly when the thermal gradient between the surface being viewed and the internal heat source is great, such as for internal fault connections in oil-filled electrical equipment. A thermographer simply will not see much of a change on the surface as the internal connection changes. Surprisingly, even objects such as bolted electrical connections often have a large gradient, even over a small physical distance. Therefore, care should always be taken when interpreting a thermal image to understand what internal conditions may be like.

A similar decrease in value occurs when external influences on surface temperature are significant or unknown. For example, this can occur when viewing the low-slope roof of a building for moisture intrusion in strong wind. Evidence of moisture cannot be seen. The characteristic thermal signature often disappears. Wet surfaces can also be confusing when either evaporation or freezing is occurring.

Spatial and Measurement Resolution

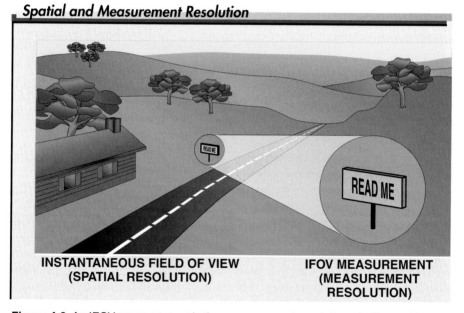

INSTANTANEOUS FIELD OF VIEW (SPATIAL RESOLUTION)

IFOV MEASUREMENT (MEASUREMENT RESOLUTION)

Figure 4-8. An IFOV measurement is the measurement resolution of a thermal imager that describes the smallest size object that can be measured at a given distance. IFOV is similar to seeing a sign in the distance while IFOV measurement is similar to reading the sign, either because it is closer or larger.

COLOR THERMAL IMAGES OF APPLICATIONS

5

Figure 5-1. The "hot spot" on the thermal image does not always indicate the primary problem. The top fuse may be blown, and the center fuse may have a possible problem as well.

Figure 5-2. Under the proper conditions, liquid level in a storage tank can be easily detected.

Figure 5-3. A blue (or dark) spot on the thermal image shows an area of unexpected moisture in a ceiling.

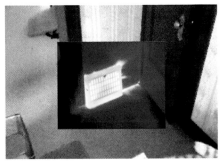

Figure 5-4. The lighter-colored, wispy thermal patterns in this picture-in-picture (PIP) thermal image of an HVAC register indicate excessive air leakage in the ductwork connections.

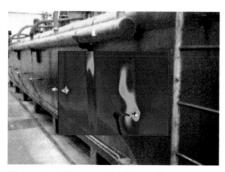 

Figure 5-5. Unexpected thermal patterns on an annealing furnace can be an indication of possible refractory insulation breakdown.

Figure 5-6. The circulation fan motor on the right side of this annealing furnace may have a potential problem, as it is operating hotter than the others.

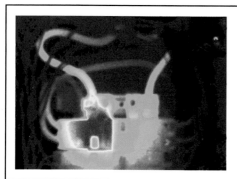 

Figure 5-7. A high-resistance connection or component malfunction within a residential circuit breaker is easily seen in the thermal image, but not in the visible light image.

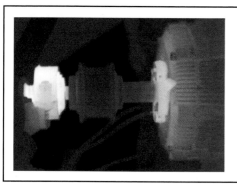 

Figure 5-8. While the thermal image of a motor and coupling shows thermal patterns on both sides that are indicative of a coupling alignment problem, the visible light image indicates no evidence of a problem.

Figure 5-9. Thermography can be used to monitor refractory performance over time and detect problem areas in cement kilns and other process equipment.

Figure 5-10. Thermography can be used to view hidden building construction and other features, such as an earthen berm on the exterior of this gymnasium.

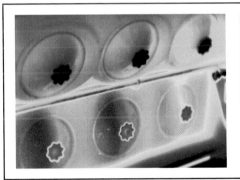

Figure 5-11. A misfire in a cylinder of a diesel-electric power plant shows different, cooler thermal patterns than normally operating cylinders.

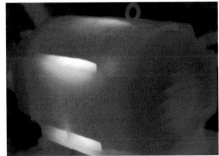

Figure 5-12. Thermal imagers can be used to scan large buildings and facilities to locate unexpected thermal variations that could indicate potential problems.

Figure 5-13. The thermal image of a normally operating motor on an air-handling system shows heat dissipating from the vents.

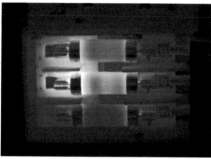

Figure 5-14. The light-colored area of the fuse bank indicates the possibility of a high-resistance issue or an internal problem with the center phase.

Figure 5-15. A hot bushing and tap on an elevated transformer is a clear sign of a problem.

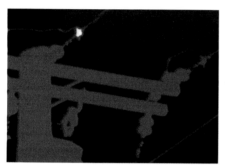

Figure 5-16. A high-resistance connection on a jumper (possibly due to corrosion) can have significant consequences if the load were to increase further.

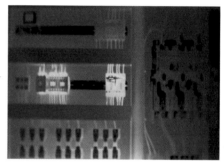

Figure 5-17. By using a thermal imager, a problem with an internal component in a motor control center (MCC) can be easily detected.

Figure 5-18. A possible load imbalance on the far right fuse can be overlooked unless level and span of the image is adjusted.

Figure 5-19. With the proper knowledge of mechanical equipment, a technician can often perform many troubleshooting and maintenance activities.

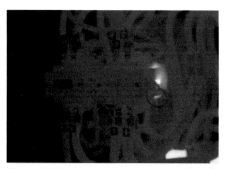

Figure 5-20. A potential internal problem is visible when comparing similar components under similar load conditions.

Figure 5-21. Thermal imagers can be used to detect wet insulation associated with a water leak on a low-sloped roof. If conditions are right and the metal roof deck is painted, it may be possible to detect such signatures from the interior.

Figure 5-22. The use of saturation colors and color alarms in a grayscale palette can be useful for determining the hot water and steam valves that are open and operating properly.

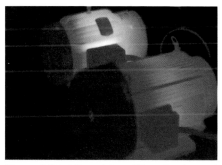

Figure 5-23. Although two different motor and pump sets show different thermal patterns, both patterns may indicate acceptable operation.

Figure 5-24. Thermal patterns in block wall construction show the intrusion of moisture at the connection of two walls, as well as unexpected construction irregularities.

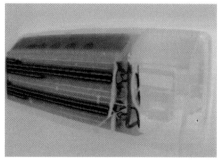

Figure 5-25. The dark-colored areas show coolant traveling through the coils of a commercial window-unit air conditioner.

Figure 5-26. The effects of low-emissivity materials on thermal imaging are apparent in the image of a metal-clad tanker truck. The metal reflects the coolness of the clear sky and the heat radiated from the ground on a sunny day.

Figure 5-27. Thermal imaging can be used to determine when equipment is not properly operating. The thermal image of the motor and pump set in the back indicates that it is unexpectedly not running.

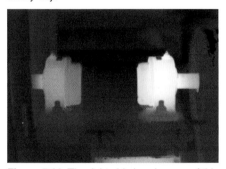

Figure 5-28. The right-side bearing cap of this air-handling unit is significantly warmer than the other indicating a potential lubrication, alignment, or belt problem.

Figure 5-29. Thermography can even be used for applications such as troubleshooting a heating wire on an outdoor water line, which cannot be allowed to freeze in cold weather.

Figure 5-30. All things on earth emit infrared energy including cold glaciers on mountaintops.

Figure 5-31. A nighttime thermal image of a container ship shows that the exhaust stack and engine room can be detected even from long distances.

Figure 5-32. In the visible light photo, it is difficult to clearly see the details of the city skyline, or the sky, on a hazy, late-summer day. However, with thermal imaging, the details, as well as the different types of clouds in the sky, can be easily seen.

 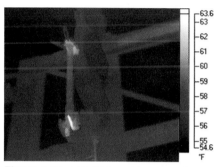

Figure 5-33. Even minor detected surface temperatures can indicate serious problems such as a shared neutral line or improper neutral-ground on a lighting system. This can cause the metal conduit inside the wall to heat up to the point where a fire hazard may exist.

Figure 5-34. Locating problems, such as a hot latch and hinge ends of a high-voltage disconnect, can be simple in conditions such as adequate load and little or no wind.

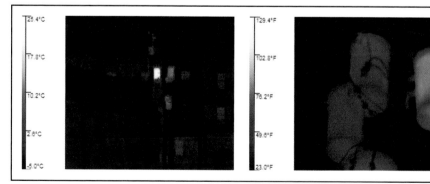

Figure 5-35. Certain problems can be detected even at a great distance (left) by using a thermal imager. More detailed analysis often requires working with a telephoto lens or by simply moving closer (right) to the equipment.

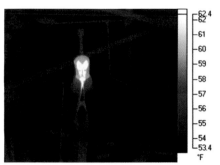

Figure 5-36. Abnormally high-resistance heating in a disconnect switch often represents a serious and costly problem because even relatively low temperatures can cause damage.

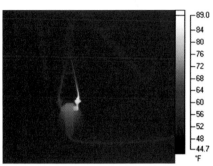

Figure 5-37. Because parallel current-flow pathways exist in many disconnect switches, the "hot spot" may be the signature of a normal connection while the cooler side may indicate the actual problem area.

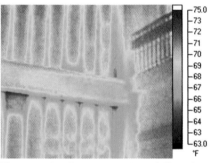

Figure 5-38. A small section of fiberglass insulation missing in a building can cause abnormal air leakage along the edges of other areas.

Figure 5-39. Because of poor air sealing, warm air can bypass the fiberglass insulation as it has in many sections of this commercial building.

Figure 5-40. A transformer that appears to be operating warmer than others on an elevated rack can be the sign of a potential problem.

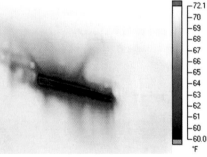

Figure 5-41. Conditioned air can leak through the joints of the HVAC ductwork into the wall behind a diffuser.

Figure 5-42. The warm areas on a surface of a boiler can be caused by refractory breakdown, air leakage, or a combination of both.

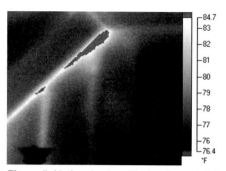

Figure 5-43. A red-colored "saturation palette" clearly shows the poor fit of fiberglass insulation in a slant ceiling.

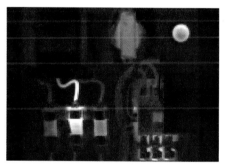

Figure 5-44. The connection to the fuse clip in a motor control center (MCC) is abnormally warm.

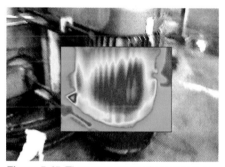

Figure 5-45. The temperature of a motor casing can quickly be checked to determine if it is operating normally.

G. McIntosh

Figure 5-46. This multi-stage air compressor is working well as can be seen by the temperature increase at each stage.

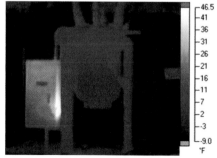

Figure 5-47. Thermography can be used to document if a heater inside a control cabinet is functioning normally to minimize problems with condensation.

Figure 5-48. A thermal image of the exterior view of a building can clearly show problem areas such as the lighter sections where insulation is missing.

Figure 5-49. Areas of missing insulation show up as warm spots from the exterior of a building during cold weather.

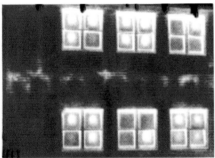

Figure 5-50. Warm spots in the center of double-glazed windows can indicate a loss of the insulating argon gas that normally fills the space between the windowpanes.

Figure 5-51. Thermal imaging can be used to document problems such as missing or damaged insulation.

Figure 5-52. An abnormally warm bearing in an overhead trolley can lead to excessive power consumption and a stretched chain over time.

Figure 5-53. An area of wet roof insulation shows up as a warm spot on the roof in the early evening when the conditions are optimal for taking thermal images.

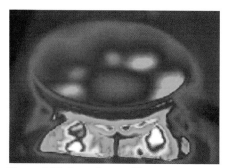

Figure 5-54. A cast iron cooking pan displays a unique thermal signature as it is heated.

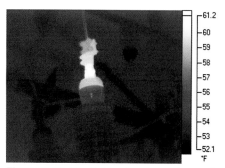

Figure 5-55. The light-colored area in an oil-filled circuit breaker shows that the internal connection from the bushing cap to the bushing rod is abnormally hot.

Figure 5-56. A light-colored area indicates the water level in a municipal water-storage tank.

Figure 5-57. When using a thermal imager, the level of liquid propane in a storage tank can easily be seen.

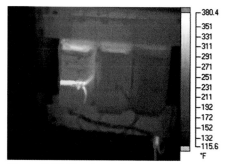

Figure 5-58. The thermal signature of a three-phase dry transformer indicates that the primary lead to the left phase is abnormally warm.

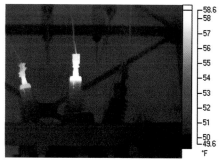

Figure 5-59. Two of the six bushing caps on an oil-filled circuit breaker are abnormally hot and represent a condition that would have proved costly had it not been detected and repaired.

Figure 5-60. A properly working steam trap should be warmer on the steam side and cooler on the condensate side as is the case in this image.

Figure 5-61. Liquid levels in storage tanks are easily seen with a thermal imager when conditions are optimal.

Figure 5-62. Although a stove burner appears warm, the flame is barely visible in a long-wave thermal image.

Figure 5-63. The many warm areas on the exterior front wall of a commercial building are associated with poorly installed fiberglass insulation.

Figure 5-64. In addition to wet insulation, many objects on a roof can have warm signatures such as the vent hood on the HVAC system.

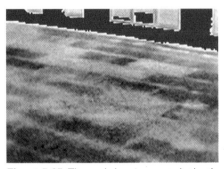

Figure 5-65. Thermal signatures on single-ply roofs with foam insulation can be more subtle than signatures on built-up roofs.

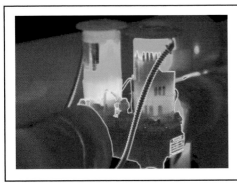

Figure 5-66. A thermal signature can be used to determine the operation of each stage in a two-stage pump.

Figure 5-67. A belt rubbing against an overhead conveyor system tray creates a hot spot in the thermal signature. The belt became misaligned due to a nearby worn roller bearing. As a result, increased friction contributed to the overheating of a driver motor.

Figure 5-68. Excessive heat loss can be caused by warm air bypassing insulation and can be a significant and costly problem in many buildings even when they are insulated.

Figure 5-69. Poorly installed loose-fill insulation in an existing wall cavity can settle and not perform as well as it should.

Figure 5-70. The nose of a person's face is often colder than other parts of the body due to less blood flow and more convective cooling.

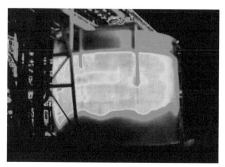

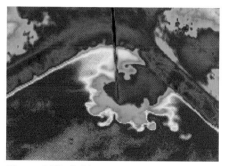

Figure 5-71. Both liquid and sludge levels in a tank are often detectable when a tank is in thermal transition.

Figure 5-72. Cold water streaming into a sink full of warm water results in convective heat transfer.

Figure 5-73. The gold-plated dome of a government building reflects the relatively cold sky.

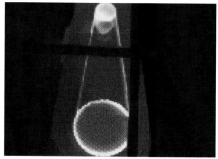

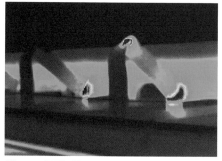

Figure 5-74. The light-colored areas on the thermal image of a belt and pulleys probably indicates an out of alignment condition.

Figure 5-75. The red-colored areas of the image indicate several conveyor roller bearings that are abnormally heated.

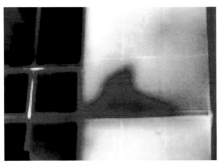

Figure 5-76. Moisture can penetrate the stone façade of a commercial structure, leaving it vulnerable to damage.

Figure 5-77. This thermal image shows a normally operating, open hydraulic valve.

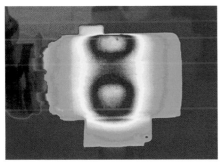

Figure 5-78. The heating pattern on a normally operating pump motor has a uniform thermal signature.

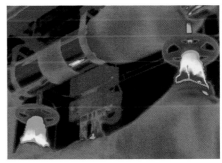

Figure 5-79. Light-colored areas indicate where heat escapes from uninsulated sections of a steam system near the valves.

Figure 5-80. Cool air that escapes from underneath an entry door leaves a wispy, finger-like pattern on the hallway floor.

Figure 5-81. The right-hand electrical plug for a bank of computer servers shows a thermal pattern that indicates either a high-resistance connection or an internal wiring problem.

Figure 5-82. The bright-colored area indicates a possible high-resistance connection or component failure in a lighting control panel.

Figure 5-83. The differing color on both sides of this in-line condenser unit and bypass value indicates expected operation.

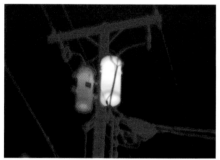

Figure 5-84. This thermal image indicates that the transformer on the right may have an internal fault.

Figure 5-85. A thermal image of a normally functioning HVAC compressor can show a wide temperature difference between different sections and components.

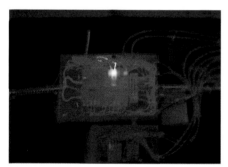

Figure 5-86. Thermal imaging can be useful in heat tracing for high-resistance connections in low-voltage control systems.

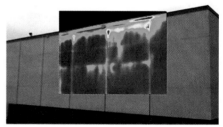

Figure 5-87. Moisture from an inadequately drained roof can leak into the concrete block and façade of a building.

THERMOGRAPHY APPLICATIONS

*T*hermography and thermal imaging can be used for applications such as inspecting electrical and process equipment and for building diagnostics. Electrical equipment includes motors, distribution equipment, and substations. Process equipment includes automated manufacturing and assembly equipment. Building diagnostics include checking for roof moisture, building insulation inspections for air leakage and moisture detection. Insulation includes material located in walls, ceilings, and floors of the thermal building envelope.

ELECTRICAL APPLICATIONS

Thermal imagers are most commonly used for inspecting the integrity of electrical systems because test procedures are noncontact and can be performed quickly. Most electrical thermography work is qualitative, that is, it simply compares the thermal signatures of similar components. A thermal signature is a snapshot at a single point in time of the heat being given off, or emitted, from an object. It is straightforward with three-phase electrical systems because under normal conditions the phases almost always have easily understood thermal signatures.

Thermography is particularly effective because equipment failures often have clear, recognizable thermal signatures. Furthermore, thermal exceptions appear even where visible inspection shows very little, if anything. A *thermal exception* is an abnormal or suspect condition that is present in equipment. Although thermal exceptions may not always be detectable, or the root cause well understood, there is no doubt that the heat produced from high electrical resistance typically precedes electrical failures.

When one or more phases or components have a different temperature, due to issues unrelated to normal load balance, a thermal exception may be present. For example, abnormally high resistance results in heating at a connection point. However, in a failed state, and thus not energized, components may appear cooler.

An open electrical enclosure can expose a thermographer to various hazards. Electrocution is typically not an issue because thermography does not require contact. However, the potential of an arc blast is higher, especially at 480 V and above.

For example, opening a door can trigger an arc flash if the latch is defective or objects, such as pests or dust and debris, inside the enclosure are disturbed. This can result in a phase-to-ground arc being established. Once started, an arc can reach temperatures in excess of 16,650°C (30,000°F) in less than half a second. Only authorized and trained personnel should open enclosures containing energized electrical equipment.

Thermographers should make every effort to understand and minimize the risk of an arc blast. International governing bodies can provide detailed requirements necessary to minimize arc-flash hazards. These requirements include education about risks, procedures for conducting inspections, and discussion of necessary personal protective equipment (PPE). PPE is designed to mitigate the potential damage caused by the intense heat of an arc flash and typically includes protection for the eyes, head, skin, and hands. **See Figure 6-1.**

Techniques for inspecting electrical systems are based on common sense, technology, and good maintenance practices. Whenever possible, components and equipment should be energized and viewed directly with a thermal imager.

Occasionally, inspections must also be performed from an indirect view such as an enclosed motor terminal box or an enclosed overhead busway. While this may be a necessary alternative in some situations, such as for an overhead busway, it is not recommended for regular procedure. If an enclosure cannot be opened, data from the thermal inspection may not provide the necessary details on its own.

Some equipment may be so difficult and/or dangerous to access that other inspection measures are required. Additional inspection methods may include the use of a viewport or an infrared transparent window to provide a view of the enclosure interior. Other technologies, such as an airborne ultrasound, can also be used.

Personal Protective Equipment (PPE)

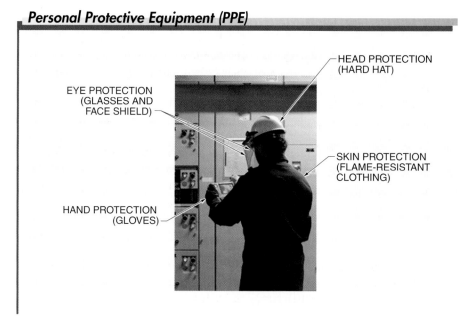

Figure 6-1. PPE typically includes eye, head, skin, and hand protection designed to mitigate the potential damage caused by the intense heat and other hazards of an arc flash.

The careful placement of an infrared transparent window is required to ensure that all components and devices can be seen. An *infrared transparent window* is a device installed in electrical enclosures to provide for the transmission of infrared energy to be viewed by a thermal imager. Infrared transparent windows can often allow thermal imaging without the need to open enclosure doors or panels. **See Figure 6-2.**

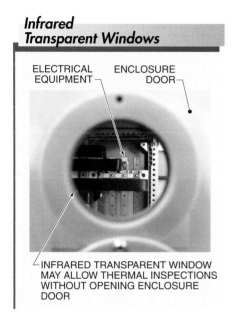

Figure 6-2. An infrared transparent window is used to provide for the transmission of infrared energy to a thermal imager without the need to open enclosure doors or panels.

It may also be possible to use equipment that detects airborne ultrasound. *Airborne ultrasound* is sound produced by a failing electrical connection. It is out of the natural hearing range but can be detected with special listening devices. Even micro arcing in a connection typically produces a detectable airborne-ultrasound signature through a crack or small hole in the enclosure.

During an inspection, particular attention is given to any electrical connection or point of electrical contact. Electrical connections and points of contact are susceptible to heat caused by abnormally high resistance and is the primary source of system failures.

Electrical current imbalances among phases can also be detected. Often these are considered normal such as in a lighting circuit. However, they can result in very costly failures in other parts of an electrical system such as an electric motor that has lost a phase or for any circuit that is overloaded.

Although thermal imagers are widely used for electrical applications, they are too often used ineffectively and improperly. Potential problems can be overlooked or, when located, misunderstood by the thermographer. Many factors, other than the severity of the problem, can influence the surface temperature seen through a thermal imaging system. Furthermore, the relationship between heat and failure, especially over time, is not always well understood.

It is well known that the temperature of an electrical connection varies as the load changes. Heat output from a high-resistance connection is predictable (I^2R) but the temperature it can achieve is much less predictable. For this reason some standards recommend that inspections be performed with a 40% minimum load or at the highest normal load when possible. Special concern should be given to any irregularities found on lightly loaded equipment where the load is likely to increase with future use.

When enclosures cannot be opened easily and heating components viewed directly, such as an overhead, enclosed busway, the thermal gradient between the problem and the viewed surface will typically be very large. A *thermal gradient* is the difference between the actual temperature at the source of a problem and the temperature that is being detected or measured on the viewed surface of the thermal imager. A surface thermal signature as small as 2.8°C (5°F) on an enclosed busway may indicate an internal failure. Oil-filled devices, such as transformers, exhibit similar or even greater thermal gradients.

To reduce unwanted glare on the display, detachable sun visors are available for use with thermal imagers.

Care must be taken for outdoor inspections when wind speed is greater than 8 kph (5 mph). For example, hotspots on equipment should be compared with how they might appear if the wind is calm. Some irregularities can cool below the threshold of detection until wind speed drops. Similar influences can occur inside a facility when enclosures are left open for a period of time prior to conducting the inspection. Good inspection procedures require that the inspection be conducted as quickly and safely as possible after opening an enclosure.

Viewing an image on a display screen while outdoors can often be challenging as well. Lighting conditions may produce unwanted glare, reducing the ability to effectively see every detail and nuance being captured. Outdoor-equipment inspections do not necessarily need to be conducted at night but bright, sunny weather can also result in confusing images due to solar heating. This is especially true for dark-colored components such as ceramic power line insulators.

The task of acquiring reliable thermal data about an electrical system is not always as simple as it seems. Even with good thermal data, many thermographers use it incorrectly when prioritizing the seriousness of the test results. For example, temperature is often not a reliable indicator of the severity of a problem because many factors can cause it to change. This fact does not prevent many thermographers from incorrectly believing that the warmer a problem component is, the more serious the problem compared to other, cooler components.

Similarly, there can be a mistaken belief that a problem does not exist when a component or piece of equipment is not particularly hot. Care must be taken in gathering and interpreting thermal data to achieve the full benefit of thermography.

Rather than determining priorities strictly on temperature, a more useful approach is to consider how all parameters interact with and affect the problem component. This can be done simply with test instruments or more formally by root-cause failure analysis using engineering analysis tools. The benefits of properly conducting thermal electrical inspections are great and companies that are successful are able to virtually eliminate unscheduled downtime due to electrical failures.

ELECTROMECHANICAL AND MECHANICAL APPLICATIONS

Electromechanical and mechanical inspections cover a diverse variety of equipment. Thermal imaging has proven invaluable for inspecting equipment such as motors, rotating equipment, and steam traps. Most of these applications are qualitative. The current thermal image is typically compared to a previous one. Any differences resulting from a change in equipment condition are then noted. A thermographer must have a solid knowledge of heat transfer in order to understand how equipment functions and fails.

Motors are thermally inspected because they are very susceptible to heat-related failure. For example, motor misalignment or imbalance typically results in overheating. While it is useful to look at the surface temperature of a motor housing, changes to the internal temperatures of a motor are not always immediately obvious. It can be valuable to take thermal images of the motor either over time or in comparison with similar operating motors. For example, this can help to reveal a motor that has become clogged with dust or that is single-phased and overheating.

The thermal signature of motor bearings can also be used for inspections. For example, if motor bearings are much warmer than the motor, it is an indication of a possible problem that should be investigated in greater detail. Similarly, motor couplings and shaft bearings that are operating normally should exhibit thermal signatures that are very close to the ambient air temperature. **See Figure 6-3.** It is useful to employ other types of testing, such as vibration or motor circuit analysis, in conjunction with thermography.

Thermal Signatures

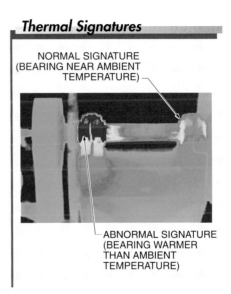

NORMAL SIGNATURE (BEARING NEAR AMBIENT TEMPERATURE)

ABNORMAL SIGNATURE (BEARING WARMER THAN AMBIENT TEMPERATURE)

Figure 6-3. Motor couplings and shaft bearings operating normally should exhibit a thermal signature that is very close to ambient air temperature.

Thermography has proven particularly valuable for inspecting low-speed rotating equipment, such as conveyors, where other inspection methods may not be useful or reliable. More complex types of equipment, such as turbines, gearboxes, and heat exchangers, can also be inspected with a thermal imager. However, they often require a more significant investment in creating a baseline of test data before the results from subsequent inspections pay any returns.

PROCESS APPLICATIONS

Thermal inspections are commonly used to monitor equipment capable of enduring high-temperatures, that is, refractory equipment. For example, maintenance technicians are able to use the thermal data to validate the condition of insulation or calculate surface temperatures that could cause a failure.

A *baseline inspection* is an inspection intended to establish a reference point of equipment operating under normal operating conditions and good working order. A *trending inspection* is an inspection performed after a baseline inspection to provide images for comparison. Monitoring trends over time often provides diagnostic and predictive information. This allows the thermographer to compare any differences or similarities that may be indicative of equipment performance.

Baseline inspections should be conducted first, followed by ongoing trending inspections. These inspections should be scheduled on a frequency determined by consequence of failure and the condition of the equipment. As a result of monitoring trends, proactive maintenance capabilities are greatly enhanced and the occurrences of unplanned downtime and expensive failures are reduced.

All types of thermal insulation can be inspected by looking for variations in the surface thermal signature. These insulation types include what is used on steam lines, product lines, and piping systems, and heat tracing on process lines (both steam and electric). Unfortunately, many types of insulation systems are often covered with unpainted metal cladding that can decrease much of the value of thermography. Thermal signatures are not readily evident on unpainted metal cladding due to its low emissivity and high reflectance.

One of the most common applications for thermography is locating or confirming the levels of solids, liquids, or gases in vessels such as storage tanks and silos. **See Figure 6-4.** Although most vessels typically have instrumentation to indicate the level of material inside them, the data is often inaccurate because the instrumentation does not function properly or at other times, the data is accurate but must be independently confirmed.

TECH-TIP

Steam traps and most valves will exhibit temperature differences across the devices when working properly. There are, of course, numerous types of steam traps and valves and each may have subtly different thermal signatures. Therefore, it is important to study them carefully over a period of time and understand how they normally function.

The rate at which these materials change temperatures during a transient heat flow cycle is determined by the mode by which heat is transferred and the differing thermal capacities of the solids, liquids, and gases in the tank. Gases change most quickly. For example, the sun can provide a detectable thermal change in the gas-filled portion of a large outdoor tank in a matter of minutes. Solids, liquids, and floating materials all change at different rates when pushed through a temperature cycle. Even an indoor tank can have some thermal fluctuation that can reveal a variety of levels.

Liquid Levels in Tanks

TANK

WATER LEVEL

THERMAL IMAGER DISPLAYS THE LEVEL OF WATER IN TANK BASED ON DIFFERENCES IN TEMPERATURE BETWEEN AIR AND WATER INSIDE

WATER STORAGE TANK

Figure 6-4. One of the most common applications for thermography is locating or confirming levels of material in vessels such as storage tanks and silos.

A skilled thermographer can often find levels in an uninsulated tank. Where insulation is present, it may take longer for thermal signatures to appear or some enhancement may be required. Material levels in a vessel can be enhanced by using simple active techniques such as applying heat or inducing cooling with evaporation. For example, simply spraying water briefly on a tank and waiting a few minutes for the exterior surface of the tank to change temperature is often enough to reveal several levels. Applying vertical stripe paint or a tape where levels can be read easily can modify the low emissivity of shiny metal-insulation coverings.

BUILDING DIAGNOSTICS

Thermal imaging has long been used for various applications related to residential and commercial building diagnostics. Building diagnostic applications include roof moisture inspections, building insulation inspections for energy and air leakage, and moisture detection. As with other thermography applications, knowledge of heat transfer theory and how buildings are constructed is required for success. Inspections for commercial buildings can be more complicated than for residential structures.

Roof Moisture Inspections

For a number of reasons related to design, installation, and maintenance, most low-slope roofs develop significant problems within a year or two of installation. A *low-slope roof* is a flat commercial roof with a slight pitch to drain precipitation. It is composed of a structural deck on which some type of rigid insulation and waterproof membrane is placed.

While the damage caused by an actual leak may be significant, the hidden long term damage caused by the trapped moisture is usually far more costly. Once it enters a roofing system, moisture becomes trapped and causes degradation and premature failure of the roof system. By locating and replacing wet insulation, subsurface moisture is eliminated and the life of a roof can be greatly extended beyond the expected average.

A roof moisture inspection performed with a thermal imager is nondestructive. **See Figure 6-5.** Wet insulation has a higher thermal capacitance than dry insulation. For example, after a warm and sunny day during an evening that is clear and free of wind, a roof can cool quickly. The rapid cooling of the roof leaves wet insulation warmer when compared to dry insulation.

Once these patterns are seen, large areas of the roof can be inspected quite quickly, noting any patterns indicating wet insulation. If necessary, the actual presence of moisture in a wet area can be confirmed with more traditional inspection methods, although these methods are often slow or destructive. The "inspection window" can remain open long into the night if conditions are good.

The exact thermal signature that can be seen on a thermal imager and when it can be viewed depends on the condition and type of roof insulation. Absorbent insulations typically used in low-slope roofs, such as fiberglass, wood fiber, and expanded perlite, yield clear thermal signatures. Nonabsorbent foam-board insulations, which are often used in single-ply roof systems, are more difficult to inspect. This is because little water is absorbed. Many single-ply roofs are also ballasted with a heavy layer of stone that can render a thermal signature of limited value.

Thermal signatures are also influenced by many conditions other than subsurface moisture. The roof surface must be dry or evaporation will reduce solar heating. A heavy covering of evening clouds can reduce cooling while excess wind can erase all thermal signatures.

Roof construction and physical conditions also shape thermal signatures. For example, a parapet wall facing west can re-radiate heat to the roof long into the night. Extra roofing gravel will stay warmer and previously repaired sections of the roof may appear to be different than other sections. Understanding these influences and the result they have on thermal signatures is essential for successful inspections.

Ideally, roofs are inspected shortly after they are installed to establish a baseline thermal signature. Another inspection would be warranted after any potentially damaging incident such as a heavy hailstorm, tornado, or hurricane. When inevitable leaks occur, a quick follow-up infrared inspection may help determine their exact location and provide an indication of the extent of the insulation damage.

Roof Moisture Inspections

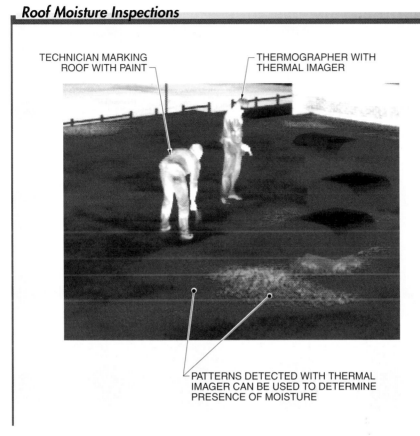

TECHNICIAN MARKING
ROOF WITH PAINT

THERMOGRAPHER WITH
THERMAL IMAGER

PATTERNS DETECTED WITH THERMAL
IMAGER CAN BE USED TO DETERMINE
PRESENCE OF MOISTURE

Figure 6-5. Roof moisture inspections are nondestructive and easily performed with a thermal imager.

Great care must be taken to conduct a roof inspection safely. Work on a roof should never be conducted alone. Thermographers are particularly vulnerable to hazards because the brightness of the display prevents their eyes from adjusting to the low-light levels found on most roofs. This is a condition referred to as night blindness. A preliminary daylight visual inspection of the roof is essential for highlighting potential hazards as well as noting roof conditions.

Building Insulation Inspections

Thermography is ideal for determining insulation presence and performance. It is being widely used by energy consultants, general contractors, and home inspectors. Insulation is mainly used within a building to control heat transfer, either gains or losses. When insulation is missing, damaged, or does not perform as intended, both energy use and the cost of conditioning increase while comfort levels in the building typically decrease.

While reducing excessive energy consumption is important, a well planned thermal inspection can also increase occupant comfort and lead to lower energy use. Other issues that can often be located through the use of thermal inspections include unwanted water leaks or moisture condensation, the buildup of ice on a roof, and the freezing of plumbing lines. Thermography also helps to check air circulation in conditioned spaces and verify the placement of acoustical insulation.

When a temperature difference of at least 10°C (18°F) is present from the inside to outside of a building, insulation problems can typically be detected. For example, during the heating season, the thermal signature for missing insulation will show as cooler on the interior and warmer on the exterior. During the cooling season, the thermal signature is the opposite. It is useful to know what type of insulation is present as each can have a unique thermal signature and time constant.

Most thermal inspections require working on both the interior and exterior of the building. However, both significant winds and direct sun may render exterior work difficult or impossible. These conditions result in effects that show up inside as well but often in more confusing ways because they are indirect. Inspections during the cooling season may be limited to the interior or to the evening for any work to be performed on the exterior. Under optimum conditions, missing, damaged, or ineffective insulation, as well as the location of framing, are often easily located by a trained, skilled thermographer properly using a thermal imager.

Air Leakage Detection

Excessive air leakage into and out of buildings accounts for nearly half the cost of heating, ventilation, and cooling. Air leakage is typically caused by pressure differences across the building. Pressure differences can be the result of wind but may also be caused by the convective stack forces inherent in any building and the pressure imbalances associated with an HVAC system.

Thermal imaging can be used to check for heat loss within a building around areas such as windows, eaves, or poorly-insulated walls.

The pressure differences force air through the many penetrations within the building. Penetrations to the thermal envelope, such as a wiring or plumbing chase, are often small and not readily evident. A *thermal envelope* is the boundary of the space that is to be heated, ventilated, or cooled within a building.

Typically, only a small temperature difference of 3°C (5°F) from inside to outside is required for detecting air leakage. The air itself cannot be seen but its temperature patterns on building surfaces often have a characteristic "wispy" thermal signature. **See Figure 6-6.** During the heating season, thermal signatures will typically be seen as cold streaks along the interior building surfaces or warm "blooms" on the outside where heated air is escaping. Air movement may also be evident inside building cavities, even interior or insulated exterior walls.

Building Surface Temperature Patterns

Figure 6-6. Temperature patterns related to air leakage often have a characteristic "wispy" thermal signature.

By artificially inducing a pressure difference on the building, air leakage patterns can be enhanced, directed, and quantified. This can be accomplished through the use of an HVAC system or a blower-door fan.

Moisture Detection

Moisture frequently finds its way into buildings to cause degradation of building materials. The point of penetration is typically a structural joint or seam, such as a failed flashing or seal. Moisture can also result from condensation. Condensation is typically caused by warm, moist air leaking from the building to the cooler building cavities. Other sources of moisture include flooding, groundwater, and leaks from plumbing and sprinkler systems.

In all these examples, the thermal signature of the moisture that is present is often clear and obvious, especially if conditions are right for the wet surface to evaporate. In this case, the surface will appear cool. Wet building materials, however, are also more conductive and, during a thermal transition, have a large thermal capacitance compared to dry ones. In this situation, the thermal signatures may not always be clear or obvious. Care should be taken to verify that conditions allow moisture to be seen if present. For example, supplemental testing with a moisture meter is recommended to confirm what is shown on the thermal image when a suspect area is detected.

Commercial Building Inspections

While the inspection of residential buildings is quite straightforward, that of large, commercial buildings is often more complicated. However, returns on investment from understanding large buildings are often significant and can usually justify a thorough inspection and analysis. It is essential that the building construction details are understood and made completely available to the thermographer in order to fully understand the complex interactions among the different building components.

Air leakage, water penetration, and condensation are the most common problems encountered in commercial buildings. A thermal imager is a powerful troubleshooting tool for the many problems encountered in a large structure. When possible, large buildings should be inspected during construction as each floor is enclosed, insulated, and relevant finishes are installed. This allows for design or construction problems to be identified and corrected before the entire building is completed and occupied.

INSPECTION METHODOLOGIES

*T*hermographers use three main methods to perform inspections with thermal imagers. These are the comparative, baseline, and trending methods. The method of choice depends on the type of equipment being inspected and the type of data that is required. Each method can be successful if used for the appropriate application.

COMPARATIVE THERMOGRAPHY

Thermographers have developed a number of methods for expanding the use of the technology. The basic method used in many thermal applications is comparative thermography. *Comparative thermography* is a process used by thermographers to compare similar components under similar conditions to assess the condition of the equipment being tested.

When comparative thermography is used appropriately and correctly, the differences between the equipment being assessed will often be indicative of the condition. Quantitative thermography, as compared to qualitative thermography, requires a more complete understanding of the variables and limitations affecting the radiometric measurement. *Quantitative thermography* is thermography that includes radiometric temperatures. *Qualitative thermography* is thermography that does not include radiometric temperatures.

It is vital to determine what margin of error is acceptable before beginning an inspection and to work carefully to stay within

those boundaries. Basic, practical training in heat transfer and advanced skills using a thermal imager are essential for understanding quantitative thermography. Much of thermography is comparative work. By comparing the object of interest to a similar object, it is often easy to detect a problem. Training and experience are fundamental to the process since there can be many other variables that must be taken into account.

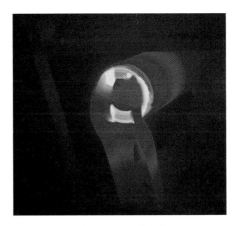

Qualitative thermography does not usually include radiometric temperatures. It compares and contrasts the thermal signatures of similar components.

For comparative thermography to be effective, the thermographer must eliminate all but one variable from the comparison. Too often this simple but essential requirement is not achieved due to the complex circumstances of an inspection or the poor work habits of the thermographer. As a result, data can be inconclusive or misleading. Care must be taken to understand the influences that result in the thermal signatures observed.

For example, a thermal image of a three-phase electrical breaker can show one phase as being warmer than the others. **See Figure 7-1.** If the loads across the three phases are balanced, uneven heating is probably associated with a high-resistance connection. However, if a load reading is taken with a digital multimeter and shows loading of 30/70/30 A, for example, the pattern is most likely related to an electrical phase imbalance.

Comparative Thermography

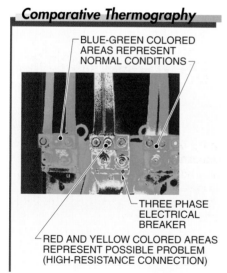

Figure 7-1. Comparative thermography can be used on a three-phase electrical breaker and show one of the phases being warmer compared to the others.

The thermal imager itself cannot "read" an image. It is a combination of the skill, experience, and persistence of a thermographer to use the system correctly, often along with other data, to achieve the correct interpretation. Of course, the misdiagnosis of an exception can result in damage to or loss of valuable equipment.

When using comparative thermography, it is useful to have as much knowledge as possible about the object being viewed. Such knowledge includes the construction, basic operation, known failure mechanism, direction of heat flow, or operating history of an object. Because this knowledge is often not readily available, the thermographer must be able to ask the equipment owner or maintenance technician clear, simple questions.

More importantly than asking questions, the thermographer must listen carefully to the answers. Many thermographers fail at either one or both of these tasks and their work suffers as a consequence. Communication skills for a thermographer are as important as technical skills, especially when working with unfamiliar equipment or materials.

BASELINE THERMOGRAPHY

A baseline inspection is intended to establish a reference point of equipment operating under normal operating conditions and in good working order. It is critical to determine what is the normal or desired equipment condition and use this as a baseline signature to which later images are compared. Often the baseline signature is uniform or related in some manner to the inherent structure of the object being viewed. For example, after a motor is installed and brought into normal operation, any differences in the thermal signature will probably show up in subsequent thermal images. **See Figure 7-2.**

Baseline Thermography

Figure 7-2. Many differences in the thermal signature of the motor will show up in subsequent thermal inspections.

THERMAL TRENDING

Another method of thermal inspection is known as thermal trending. *Thermal trending* is a process used by thermographers to compare temperature distributions in the same component over time. Thermal trending is used extensively especially for inspecting mechanical equipment where normal thermal signatures may be complex. It is also useful when the thermal signatures that detect failure often develop slowly over time. For example, thermal trending can be used when monitoring the performance of a refractory (high-temperature) insulation in a specialty

railroad car over time to determine optimum maintenance downtime scheduling. **See Figure 7-3.**

It is important for a thermographer to understand all variables present with the equipment that is inspected. Thermographers must understand the operating principles of various systems and develop troubleshooting skills. If data is carefully gathered and changes are understood, these methods can reveal a very accurate and useful trend of performance. However, it is important to remember that trending only implies, rather than predicts, the future.

Thermal Trending

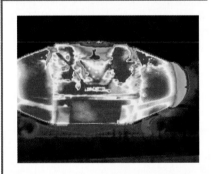

Figure 7-3. Thermal trending is used for inspecting high-temperature equipment where normal thermal signatures, such as this torpedo car (filled with molten metal), can be complex, and may only indicate insulation failure over time.

Palettes

A *palette* is a color scheme used to display the thermal variations and patterns in a thermal image. Whether inspecting or analyzing, the objective is to select the palette that best identifies and communicates the problem. Ideally, a thermal imager that allows the user to select or change the desired palette both in the camera and in the software should be chosen. For example, certain applications may be better viewed and analyzed in a monochromatic palette such as grayscale or amber. Other situations may be easier to analyze and explain in a color palette such as ironbow, blue-red, or one with a high contrast. A wide selection of available color palettes allows the thermographer greater flexibility in thermal inspection, analysis, and reporting.

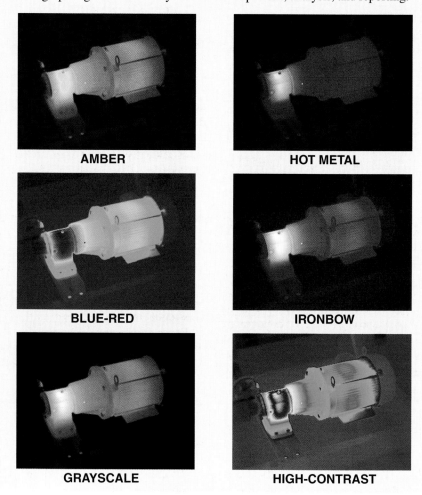

AMBER   HOT METAL

BLUE-RED   IRONBOW

GRAYSCALE   HIGH-CONTRAST

ANALYSIS, REPORTING, AND DOCUMENTATION

*I*n addition to being able to properly handle and use a thermal imager, a thermographer's job is to analyze, report, and document the results of the equipment they inspect. Special analysis tools are available to properly complete these tasks.

INSPECTION ANALYSIS

Thermography is highly dependent on the ability of a thermographer to conduct an inspection correctly, understand the limitations of the work, record all relevant data, and properly interpret the results. The variables encountered by a thermographer can be varied and numerous. As a result, thermographers must be properly trained and qualified to perform thermal inspections.

Thermographers can be certified as Level I, Level II, or Level III with Level I being the lowest level of certification and Level III the highest level. Under the implementation of a formal thermography program, a certified Level I thermographer is qualified to collect data but must work under the supervision of a certified Level II thermographer. Level II thermographers are qualified to interpret data and write reports. A formal thermography program must have written inspection procedures, typically based on industry standards and developed with the support of a certified Level III thermographer.

REPORTING AND DOCUMENTATION

After thermal data is correctly evaluated, the results may be required to be clearly communicated in the form of a written report. Part of the reporting process may require educating the customer about the inherent limitations of thermal imaging technology and the value of thermal inspections. In the end, the report often results in prescribed actions to correct all problems revealed during the thermal inspection.

A thermographer typically also provides additional information including the location of the problem, diagnosis, and suggested corrective actions. A thermographer provides key information from a thermal inspection that must be merged with additional information from other inspections or tests, maintenance or repair scheduling, and cost analysis before a successful conclusion is achieved. Therefore, good communication skills are as essential as good technical skills.

Reports can come in many styles and include a variety of data. However, a report should include the following information:

• Name of thermographer

• Make, model, and serial number of thermal imager

• Relevant ambient environmental conditions, such as wind speed, wind direction, precipitation, humidity, and ambient air temperature

- System conditions, such as load and duty cycle
- Identification and location of equipment and components inspected or tested
- List of critical equipment not inspected or tested along with explanations for the omission
- Instrument parameter settings, such as emissivity and background settings
- Thermal images and matching visual images of all equipment and components inspected or tested
- A section that requests a follow-up infrared image to document equipment repair

Documentation must also be displayed in a manner that does not clutter the report but rather supports the presentation of essential information in a clear and efficient manner. The best thermal inspection reports have a natural flow of data to support the thermal and visual images. **See Figure 8-1.**

Having access to a collection of several different report templates can be useful. For example, a simple report template can be used for documentation of successful repairs on equipment that was thermally inspected or tested. Specialized report templates can be used for particular categories of thermal inspection.

Whenever a thermography report is processed, additional copies of each report should be given to key personnel as required. Copies can be either a hard copy format or in electronic form. Electronic reports should be saved and locked (such as a PDF format) before sending them out to prevent tampering with the inspection and test analysis.

Additional value from thermal tests and inspections can often be gained by keeping track of individual problems in a more categorized and specific manner. For example, information related to issues with a specific brand of equipment or a certain process can be identified and stored. Later it can be retrieved and the issues identified as common to certain equipment to aid future users.

In addition to properly handling and using a thermal imager, a successful thermographer must be able to analyze and document the results through proper reporting. This ability is necessary in order to develop and maintain a good reputation of high-quality, consistent work. Reporting provides the best possible post-inspection recommendations.

Thermal Reporting and Documentation

VISUAL IMAGE

THERMAL IMAGE

Figure 8-1. Thermal inspection reports typically include thermal images and matching visual images for reference purposes.

THERMOGRAPHY RESOURCES

Resources can be used to gain additional information on thermography and thermal imagers such as equipment updates, safety concerns, training seminars, educational tools, and standards and professional organizations. These resources are available in electronic or printed formats.

RESOURCES

A considerable amount of relevant information concerning thermography and thermal imagers is available to most prospective users through various types of resources. In industrial-commercial and building diagnostics applications, the technology has been available for more than 40 years. However, many professionals, such as maintenance technicians and electricians, are in the early stages of learning about the technology and benefits of thermography and thermal imagers.

Resources include written material from various standards organizations.

Because of the development of new information, the application of thermography has increased rapidly in the past few years. Please be aware that some information related to thermography, especially that being published on the worldwide web, may not be accurate or factual. It is strongly suggested to learn the basics from sources such as this publication and those listed below. It is also recommended to read critically to learn more from unknown sources. Resources include standards, on-line resources, books and printed materials, and professional organizations.

Standards

A *standard* is an accepted reference or practice developed by industry professionals. Standards provide a set of acceptable criteria to which work can be performed. While compliance with an industry standard can be voluntary, it is good practice to comply with approved, recognized standards. Standards are created from the input of various industry experts and are available through various organizations. **See Figure 9-1.** They can be a valuable resource in providing specific, detailed information on various aspects of thermography.

On-line Resources

An *on-line resource* is a resource that is available to users only through an Internet connection. These resources provide a variety of instructional resources for students, thermographers, and technicians. Supplemental information is typically available from equipment manufacturers, standards organizations, learning materials, and professional organizations. For example, an on-line resource can include a live forum where experienced representatives from equipment manufacturers can "chat" with equipment users for the purpose of troubleshooting or providing equipment recommendations.

Books and Printed Materials

Books and printed materials are sources on hard copy that can be used as technical references. They serve to increase the knowledge of any individual who uses thermography and infrared technology for testing and inspection purposes. There are several books and printed materials available.

Professional Organizations

A *professional organization* is an organization that provides information and education regarding thermography through publications, training events, and participation in local chapters. Thermographers and technicians are encouraged to join and participate in various professional organizations. Such membership helps individuals to maintain an awareness of the latest in technology, trends, and changes in the industry. Participation in a professional organization

provides new opportunities and aids in gaining knowledge about thermography processes and the latest equipment and testing/inspection techniques.

Standards Organizations
American Society for Nondestructive Testing (ASNT)
1711 Arlingate Lane   www.asnt.org PO Box 28518 Columbus, OH 43228 614-274-6003
ASTM International (ASTM)
100 Barr Harbor Drive   www.astm.org PO Box C700 West Conshohocken, PA 19428 610-832-9598
Canadian Standards Association (CSA)
5060 Spectrum Way   www.csa.ca Suite 100 Mississauga, ON L4W 5N6
Institute of Electrical and Electronics Engineers (IEEE)
1828 L Street NW   www.ieee.org Suite 1202 Washington, DC 20036 202-785-0017
International Electrotechnical Commission (IEC)
3, rue de Varembe'   www.iec.ch PO Box 131 CH-121 Geneva 20 Switzerland
International Organization for Standardization (ISO)
1, ch. de la Voie-Creuse   www.iso.org Case postale 56 CH-1211 Geneva 20, Switzerland +41 22 749 01 11
National Fire Protection Association (NFPA)
1 Batterymarch Park   www.nfpa.org Quincy, MA 02169 617-770-3000

Figure 9-1. Standards are accepted references or practices developed by industry professionals and are available through various organizations.

OTHER RELATED TECHNOLOGIES

*I*n addition to thermography, other related technology and methods for analysis are used to inspect and troubleshoot commercial and industrial equipment and components. These methods include visual and auditory inspection, electrical analysis, ultrasonic analysis, vibration analysis, lubricating oil analysis, and wear particle analysis. They can be used alone to troubleshoot equipment or following the use of thermal imagers to verify the test results obtained.

VISUAL AND AUDITORY INSPECTION

Visual and auditory inspection is the analysis of the appearance of problems and sounds of operating equipment to determine components that may need maintenance procedures or repair work performed. **See Figure 10-1.** Visual and auditory inspection is the simplest PdM procedure performed in a facility and requires no tools or equipment. It is most effective where a potential problem is obvious to the trained maintenance technician. Extraordinary operating characteristics are noted and the equipment is scheduled for the required maintenance.

Visual inspection can be supplemented with processes such as dye-penetrant testing to locate fine, surface metal fractures. The metal is completely cleaned and sprayed with dye that collects in small fractures or pits on the surface of the metal. Excess dye is removed to reveal small cracks or pits below the surface.

Visual And Auditory Inspection

Figure 10-1. Maintenance technicians routinely check the appearance and sounds of operating equipment by conducting visual and auditory inspections.

ELECTRICAL ANALYSIS

Electrical analysis is a method of analysis that uses electrical monitoring equipment to evaluate the quality of electrical power delivered to equipment and the performance of electrical equipment. **See Figure 10-2.** Electrical monitoring equipment can be installed to measure minimum and maximum voltages, phase-to-phase voltage variation, loss of voltage, and current levels. It can also assess the quality of power being supplied to sensitive electronic equipment.

Electrical Analysis

Figure 10-2. Electrical analysis uses electrical monitoring equipment to evaluate the quality of power delivered to equipment.

One of the most common applications of electrical analysis is used for electric motors and circuits. *Motor circuit analysis (MCA)* is a type of electrical analysis for motors and circuits that can be conducted either on-line (energized) or off-line (de-energized). Both testing methods provide early detection of defects or faults in the electrical distribution of motors, motor circuits, and motor drive trains.

AIRBORNE ULTRASOUND DETECTION

Airborne ultrasound detection is a method of equipment analysis that amplifies high-frequency sound to identify possible equipment problems. A sensitive listening device converts these sounds, which are normally out of the range of hearing, to signals that can be perceived by humans. These signals can indicate, among other things, the abnormal heating of electrical connections, leaks in air and steam systems, and friction in bearings, as well as many other equipment problems.

VIBRATION ANALYSIS

Vibration analysis is the monitoring of individual component vibration characteristics to determine the condition of the equipment. Worn parts frequently cause equipment failure. They also produce increased vibration or noise that can be isolated. Vibration analysis is the most common form of monitoring technique that is used on rotating equipment.

LUBRICATING OIL ANALYSIS

Lubricating oil analysis is a predictive maintenance technique that detects the presence of acid, dirt, fuel, or wear particles in lubricating oil and examines those substances to predict equipment failure. Lubricating oil analysis is performed on a scheduled basis. A sample of oil is taken from a machine to determine the condition of the lubricant and moving parts. Samples are usually sent to a company specializing in lubricating oil analysis.

WEAR PARTICLE ANALYSIS

Wear particle analysis is the examination of wear particles present in lubricating oil. While lubricating oil analysis focuses on the condition of the lubricating oil, wear particle analysis concentrates on the size, frequency, shape, and composition of the particles produced from worn parts. The equipment condition is assessed by monitoring wear particles. Normal wear occurs as equipment parts are routinely in contact with each other. An increase in the frequency and size of wear particles in the lubricating oil indicates a worn part or predicts possible failure.